WORKBOOK FOR

THE TECHNOLOGY OF DRAFTING

With an Introduction to

Computer-Aided Drawing

Edward A. Maruggi

Rochester Institute of Technology

Technical illustrations by Albert F. Luiz

Merrill Publishing Company
A Bell & Howell Information Company
Columbus   Toronto   London   Melbourne

Cover Photo: Larry Hamill

Published by Merrill Publishing Company

A Bell & Howell Information Company

Columbus, Ohio 43216

Copyright © 1989, by Merrill Publishing Company. All rights reserved. No part of this book may be reproduced in any form, electronic or mechanical, including photocopy, recording, or any information storage and retrieval system, without permission in writing from the publisher. "Merrill Publishing Company" and "Merrill" are registered trademarks of Merrill Publishing Company.

International Standard Book Number: 0-675-21234-0

Printed in the United States of America

1 2 3 4 5 6 7 8 9--92 91 90 89

Preface

This workbook of practice exercises has been specifically designed for use with the textbook <u>The Technology of Drafting</u> published by Merrill Publishing Company. Its purpose is to help students obtain technical drafting skills through the use of both manual and computer-aided tools and materials. The exercises have been ordered to allow the student to progress from basic to more advanced drawing problems. The exercises are sequentially coordinated with the sections of the text and reinforce the development of technical drafting techniques, technical content, knowledge, and standard practices outlined in each section. In addition, clearly stated objectives are used as drawing instructions for the student.

The practice exercises may be used in a self-paced mode, as laboratory assignments or for homework. It is recommended that no assumption be made as to the student's level of drafting skill. Emphasis should be placed on beginning skills that include appropriate line weight, use of tools and equipment, lettering, geometric constructions, and the development of forms, shapes, and figures. Tolerancing and dimensioning should be worked in as necessary after basic skill development has occurred.

The following is a list of recommended equipment and tools that the student may require to complete the practice exercises in the workbook.

1. 30/60-degree triangle
2. 45-degree triangle
3. Adjustable triangle
4. Bow compass
5. Vellum drawing paper
6. Plain bond paper
7. Reproducible or nonreproducible grid paper of 4, 8, or 12 squares to the inch
8. Ames Lettering Guide
9. T-square or parallel straightedge
10. Drawing board, 20 x 26 inches
11. Mechanical lead holder
12. H, 2H, 3H, 4H, and 6H drawing leads (regular and 0.5-mm size for metric holder)
13. Circle template
14. Ellipse template (various angles)
15. Mechanical lettering aids
16. Drafting tape
17. Erasing shield
18. Pink pearl or white S-Mars eraser
19. Dividers
20. Cleaning brush
21. 12-inch flat scale (32nds and 50ths increments)
22. Lead pointer

23. Hand-held calculator
24. 3 1/2-inch, 5 1/4-inch, and/or 8-inch floppy disk for use on micro, mini, or mainframe CAD systems
25. 12-inch metric scale
26. Irregular curves

Items 1, 2, 3, 9, and 10 in the list above may be omitted if a drafting machine is available.

TABLE OF CONTENTS

EXERCISE
IDENTIFICATION

USE OF DRAFTING EQUIPMENT AND TOOLS

Use of Instruments	2A
Use of Instruments	2B
Use of Templates and Tools	2C
Use of Curves	2D
Measurement	2E
Dual Dimensioning	2F
Dual Dimensioning	2G

LETTERING AND TEXT PRESENTATION

Guidelines and Lettering	3A
Freehand Lettering	3B
Freehand Lettering	3C
Use of Mechanical Lettering Guide	3D

THE TECHNICAL SKETCH

Sketching	4A
Circle Sketching	4B
Multiview Sketching	4C
Multiview Sketching	4D
Isometric Sketching	4E
Isometric Sketching	4F
Oblique Sketching	4G
Perspective Sketching	4H

THE GEOMETRIC CONSTRUCTION

Geometric Constructions	5A
Geometric Constructions	5B
Geometric Constructions	5C
Geometric Constructions	5D
Geometric Constructions	5E
Geometric Constructions	5F
Geometric Constructions	5G
Geometric Constructions	5H

THE MULTIVIEW DRAWING

View and Surface Identification	6A
View and Surface Identification	6B
Two-view Drawing	6C
Two-view Drawing	6D
Three-view Drawing	6E
Three-view Drawing	6F
Bracket, Rod	6G
Bracket, Clevis	6H
Clamp	6I
Six-view Drawing	6J

DIMENSIONING

Block, Holding	7A
Bracket, Guide	7B
Dimensional Conversion	7C
Applying Dimensions	7D
Dimensioning Diameters	7E
Applying Dimensions	7F
Dimensioning Chords, Arcs, and Angles	7G
Dimensioning Rounded Ends and Slots	7H
Dimensioning Round Holes	7I
Dimensioning a C Bore	7J
Dimensioning a CSK and C Drill	7K
Dimensioning a SF	7L
Dimensioning a Chamfer	7M
Dimensioning a Keyseat	7N
Support, Bearing	7O
Plate, Mounting	7P
Coupling, Shaft	7Q

TOLERANCING

Tolerances	8A
Limit Dimensions	8B
Shaft and Hole Fits	8C
Accumulation of Tolerances	8D
Feature Control Symbols	8E
Geometric Tolerancing	8F
Geometric Tolerancing	8G

THE SCREW THREAD AND FASTENERS

Screw Threads	9A
Hardware	9B
Fasteners	9C
Fasteners	9D
Fasteners	9E
Fasteners	9F
Fasteners	9G

THE SECTIONAL VIEW

Housing, Flanged	10A
Support, Shaft	10B
Coupling	10C
Bushing, Retainer	10D
Chisel, Stone	10E
Housing, Bearing	10F
Wheel	10G
Plate, Face	10H
Assembly, Shaft Coupling	10I

THE AUXILIARY VIEW

Front Auxiliary View	11A
Front Auxiliary View	11B
Top Auxiliary View	11C
Side Auxiliary View	11D
Primary and Secondary Auxiliary Views	11E
Partial Auxiliary View	11F
Curved Surface Auxiliary View	11G
Auxiliary Views	11H

THE DEVELOPMENT DRAWING

Development	12A
Development	12B
Development - Truncated Prism	12C
Development - Truncated Prism	12D
Development - Truncated Cylinder	12E
Development - Truncated Pyramid	12F
Development - Truncated Cone	12G
Development - Transition Piece	12H

DESCRIPTIVE GEOMETRY

True Length of a Line	13A
True Length and Edge View	13B
Edge View of a Plane	13C
True Size and True Shape	13D
True Distance Between a Plane Surface and a Point in Space	13E
True Angle Between Two Plane Surfaces	13F
Intersection of a Line and a Plane	13G
Intersection of Two Planes Using the Line Projection Method	13H

THE PICTORIAL DRAWING

Pictorial, Isometric	14A
Pictorial, Isometric	14B
Pictorial, Isometric	14C
Pictorial, Oblique Cabinet	14D
Pictorial, Oblique Cabinet	14E
Angular Type Perspective Drawing	14F
Angular Type Perspective Drawing	14G

THE ASSEMBLY AND DETAIL DRAWING

Bracket - Mono Detail	15A
Plumb Bob - Multi Detail	15B
Assembly, Hammer (Separable)	15C
Assembly, Detent (Inseparable)	15D
Bracket, Holding (Detail Assembly)	15E
Assembly, Injector (Expanded Assembly)	15F
Working Drawings - Gear Puller	15G
Ball, Gear Puller	15Ga
Screw, Gear Puller	15Gb
Pin, Gear Puller	15Gc
Finger, Gear Puller	15Gd
Handle, Gear Puller	15Ge
Yoke, Gear Puller	15Gf
Assembly, Gear Puller	15H
Bracket, Mounting	15I
Bracket, Offset	15J

AN INTRODUCTION TO COMPUTER-AIDED DRAWING

Format	16A
Objects	16B
Editing	16C
Template	16D
Strap	16E
Detail	16F
Cube	16G
Programming	16H
Housing, Shoe	16I
V-Block	16J
Steps	16K
Bracket, Mounting	16L
Bracket	16M
True Length of Line	16N
Wedge	16O
Truncated Pyramid Development	16P
Strap, Housing	16Q
Cover, Chassis	16R

SECTION	EXERCISE	OBJECTIVE: Using manual drafting tools redraw the two figures below at a 2:1 scale in the space provided.
2	A	

DRAWN	DATE	CHECK	SCALE 2:1	TOLERANCES XX XXX ANG	TITLE
					USE OF INSTRUMENTS

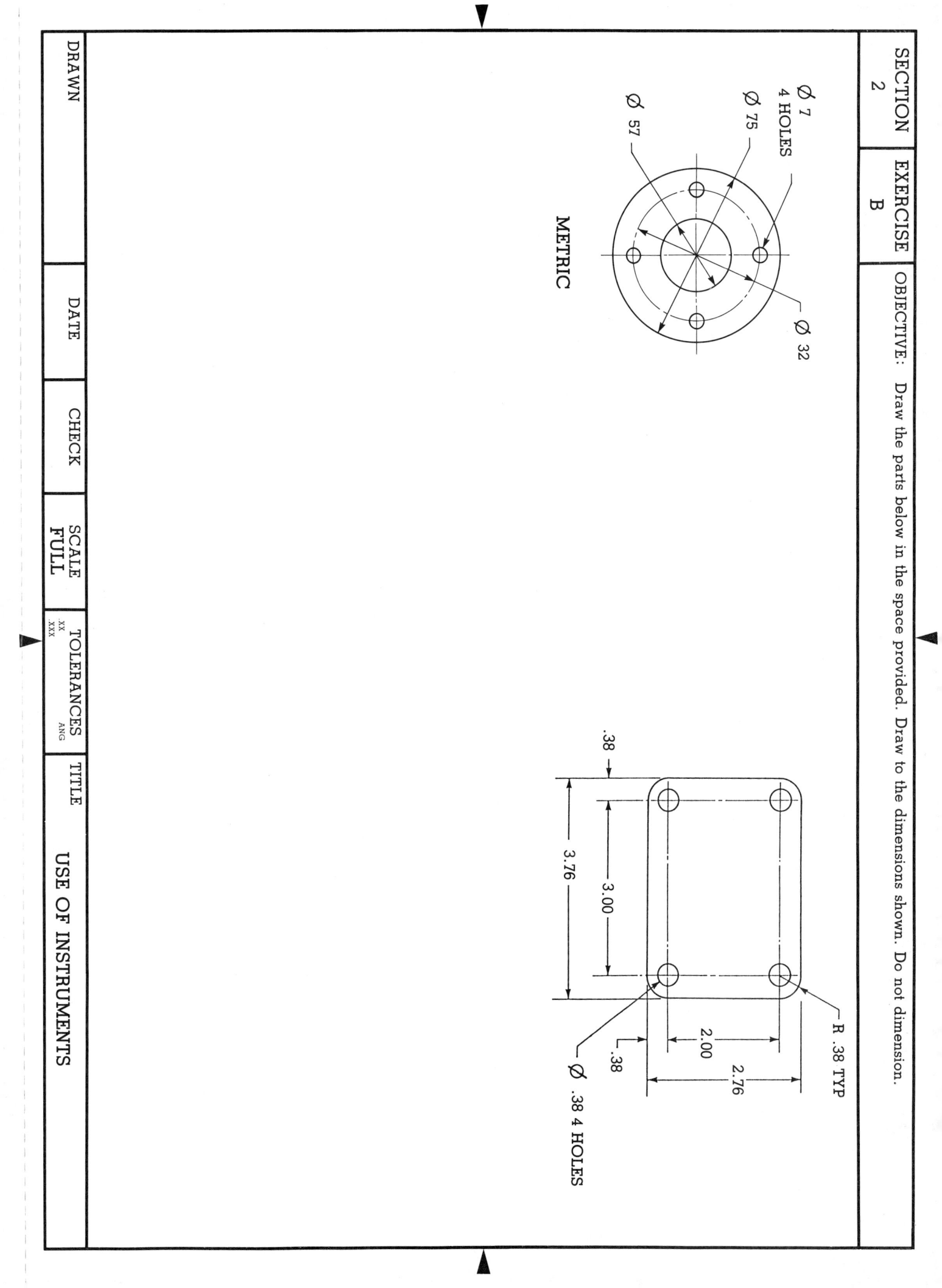

SECTION	EXERCISE	OBJECTIVE: Draw the parts below in the space provided. Draw to the dimensions shown. Do not dimension.
2	B	

Ø 7
4 HOLES

Ø 75

Ø 57

Ø 32

METRIC

.38

3.76

3.00

R .38 TYP

2.00

2.76

.38

Ø .38 4 HOLES

DRAWN	DATE	CHECK	SCALE FULL	TOLERANCES XX XXX ANG	TITLE USE OF INSTRUMENTS

SECTION	EXERCISE	OBJECTIVE: Redraw the objects shown at twice size using templates and tools. Do not dimension.
2	C	

DRAWN	DATE	CHECK	SCALE 2:1	TOLERANCES XX XXX ANG	TITLE
					USE OF TEMPLATES AND TOOLS

SECTION	EXERCISE	OBJECTIVE: Using a curve complete the figures below using the information provided in Section 2.
2	D	

DRAWN	DATE	CHECK	SCALE NONE	TOLERANCES XX / XXX / ANG	TITLE
					USE OF CURVES

SECTION	EXERCISE	OBJECTIVE: In the four areas identified below determine the length of each line to the scale indicated. Place the measurement above each line.
2	E	

METRIC SCALE
(FULL SCALE)

INCH/FOOT DECIMAL SCALE
(FULL SCALE)

ENGINEER'S SCALE

(1" = 200')

(1" = 2 MILES)

(1" = 50')

(1" = 100')

(1" = 1')

ARCHITECT'S SCALE

(1/4" = 1'-0")

(3/4" = 1'-0")

(1/8" = 1'-0")

(1/2" = 1'-0")

(3" = 1'-0")

(3/16" = 1'-0")

DRAWN	DATE	CHECK	SCALE AS REQ'D	TOLERANCES XX XXX ANG	TITLE

MEASUREMENT

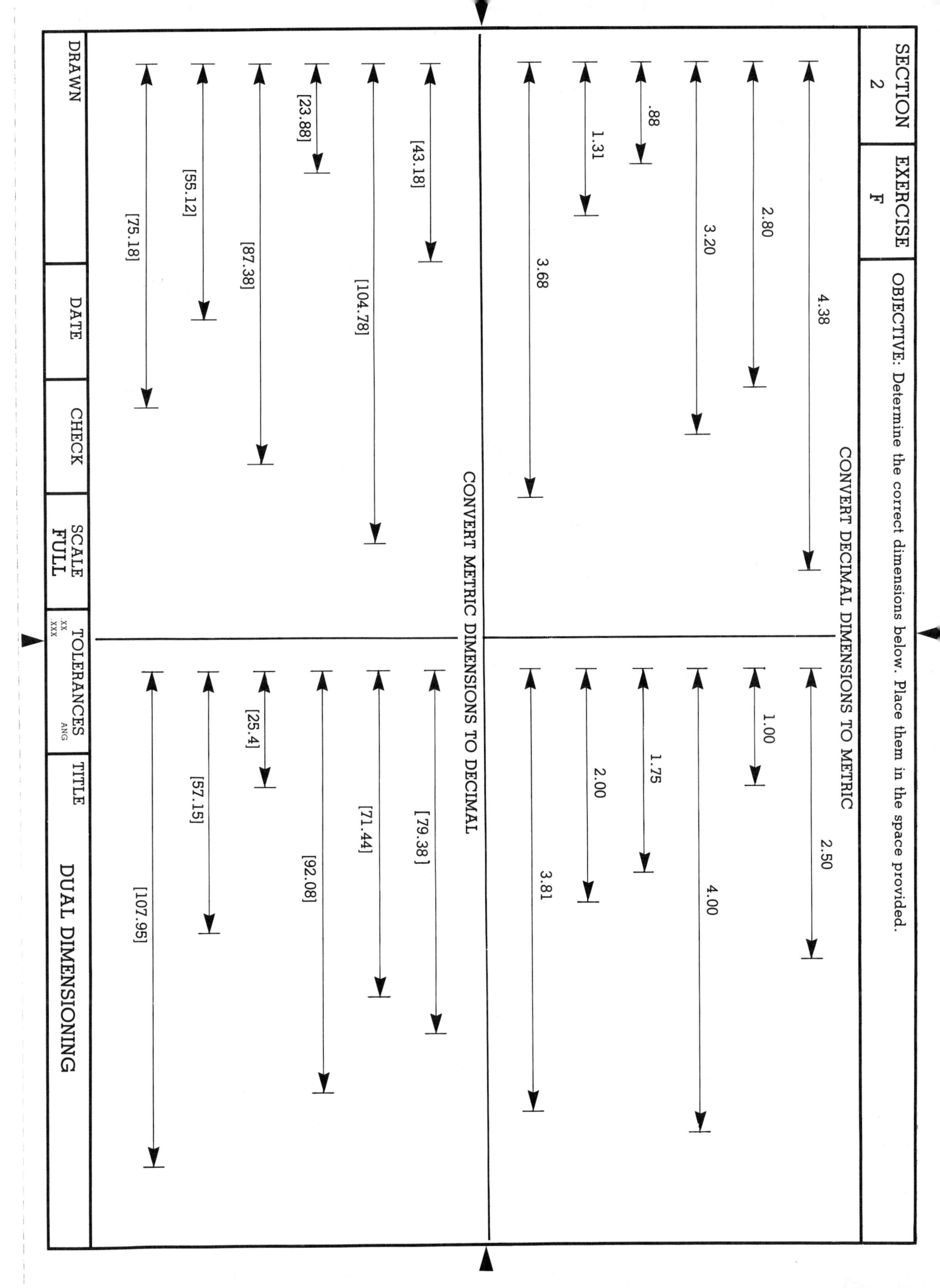

SECTION	EXERCISE	OBJECTIVE: Determine the correct dimensions below. Place them in the space provided.
2	F	

CONVERT DECIMAL DIMENSIONS TO METRIC

4.38 [ ]
2.80 [ ]
3.20 [ ]
1.31 [ ]
.88 [ ]
3.68 [ ]

1.00 [25.4]
1.75 [ ]
2.00 [ ]
2.50 [ ]
4.00 [ ]
3.81 [ ]

CONVERT METRIC DIMENSIONS TO DECIMAL

[23.88]
[43.18]
[55.12]
[75.18]
[87.38]
[104.78]

[25.4]
[57.15]
[71.44]
[79.38]
[92.08]
[107.95]

DRAWN	DATE	CHECK	SCALE FULL	TOLERANCES XX XXX ANG	TITLE
					DUAL DIMENSIONING

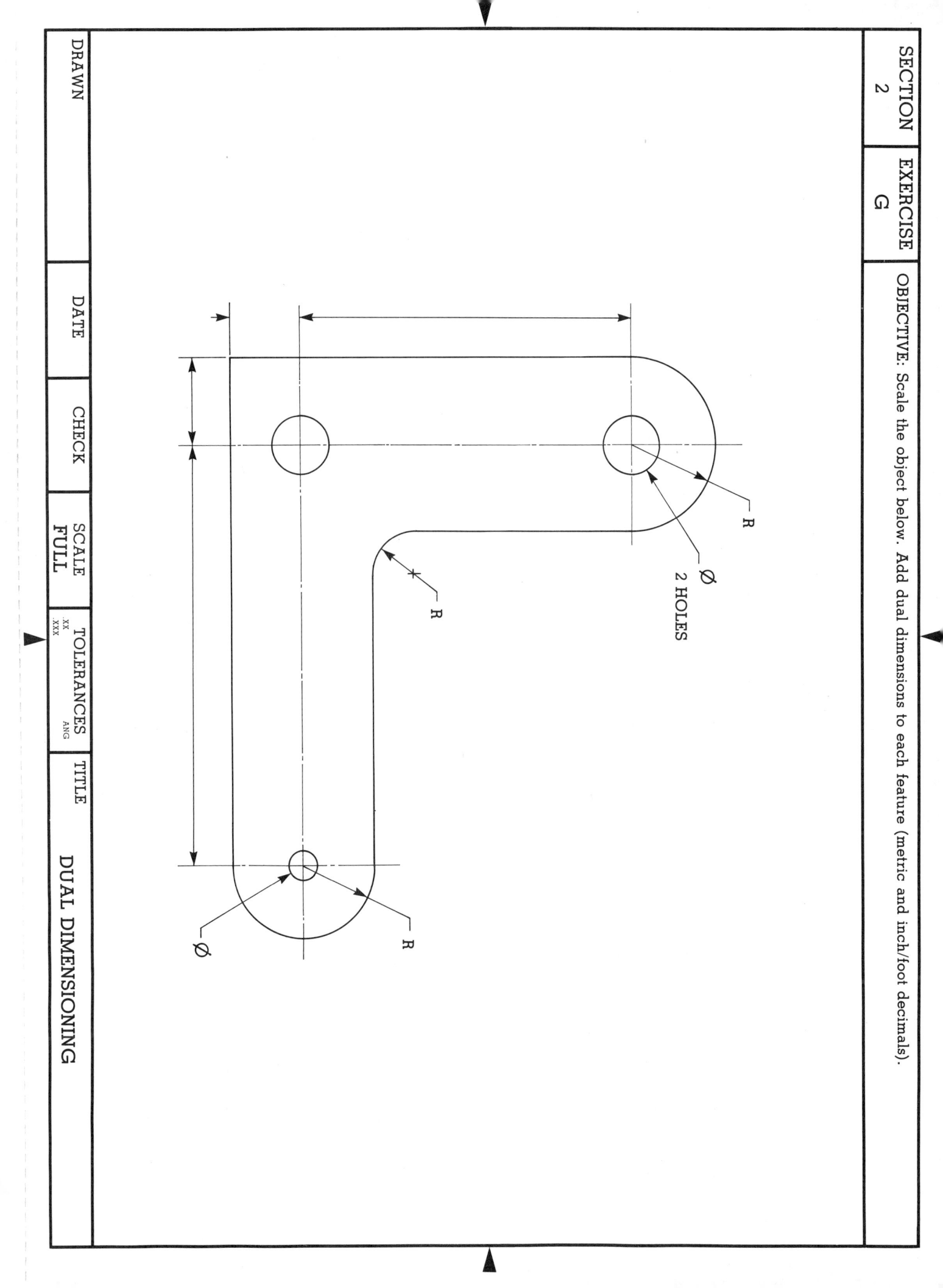

SECTION	EXERCISE	OBJECTIVE:
2	G	Scale the object below. Add dual dimensions to each feature (metric and inch/foot decimals).

R

Ø
2 HOLES

R

Ø

R

DRAWN	DATE	CHECK	SCALE FULL	TOLERANCES XX XXX ANG	TITLE
					DUAL DIMENSIONING

| SECTION 3 | EXERCISE A | OBJECTIVE: With an Ames Lettering Guide or equivalent, draw horizontal and vertical guide lines next to the indicated heights. Then letter the alphabet and numerals shown in the example using vertical gothic letters. |

EXAMPLE: ABCDEFGHIJKLMNOPQRSTUVWXYZ 1234567890

.125 HIGH

.190 HIGH

.25 HIGH

.312 HIGH

.375 HIGH

| DRAWN | DATE | CHECK | SCALE AS REQ'D | TOLERANCES .XX .XXX ANG | TITLE GUIDELINES & LETTERING |

SECTION	EXERCISE	
3	B	OBJECTIVE: Produce the lettering the text presentation below using .190 high vertical Gothic style letters. Do it twice.

QUALITY LETTERING AND TEXT PRESENTATION IS A
SKILL THAT EVERY DRAFTER SHOULD MASTER. TO
ACHIEVE THIS GOAL PRACTICE IS NECESSARY. ALWAYS
USE LIGHT WEIGHT VERTICAL AND HORIZONTAL
GUIDELINES WHEN LETTERING.

DRAWN	DATE	CHECK	SCALE AS REQ'D	TOLERANCES XX XXX ANG	TITLE FREEHAND LETTERING

SECTION	EXERCISE	OBJECTIVE: Re-produce the text below using .125 high commercial Gothic style letters. Repeat twice.
3	C	

DRAFTING TECHNOLOGY IS A METHOD OF COMMUNICATION AND IS THE LANGUAGE OF INDUSTRY. IT IS THE ART OF MECHANICAL DRAWING AS IT PERTAINS TO THE FIELD OF ENGINEERING. IN COMMUNICATING THROUGH THE MEANS OF DRAFTING TECHNOLOGY; LINES, SYMBOLS, SHAPES, VIEWS, DIMENSIONS, AND DETAILED INFORMATION TAKES THE PLACE OF MANY WORDS.

DRAWN	DATE	CHECK	SCALE AS REQ'D	TOLERANCES XX XXX ANG	TITLE
					FREEHAND LETTERING

SECTION	EXERCISE	OBJECTIVE: Using a vertical lettering guide (Mars Staedler or equivalent) re-produce the text below twice in the space
3	D	provided. Use .19 and .25 high letters.

Designing a product from initial concept to final assembly requires the energy and talent of various technical and professional people. One of the key people in the process is the drafter who, through the skill of drafting technology completes the task so that parts and assemblies can be produced.

DRAWN	DATE	CHECK	SCALE AS REQ'D	XX XXX TOLERANCES ANG	TITLE USE OF A MECHANICAL LETTERING GUIDE

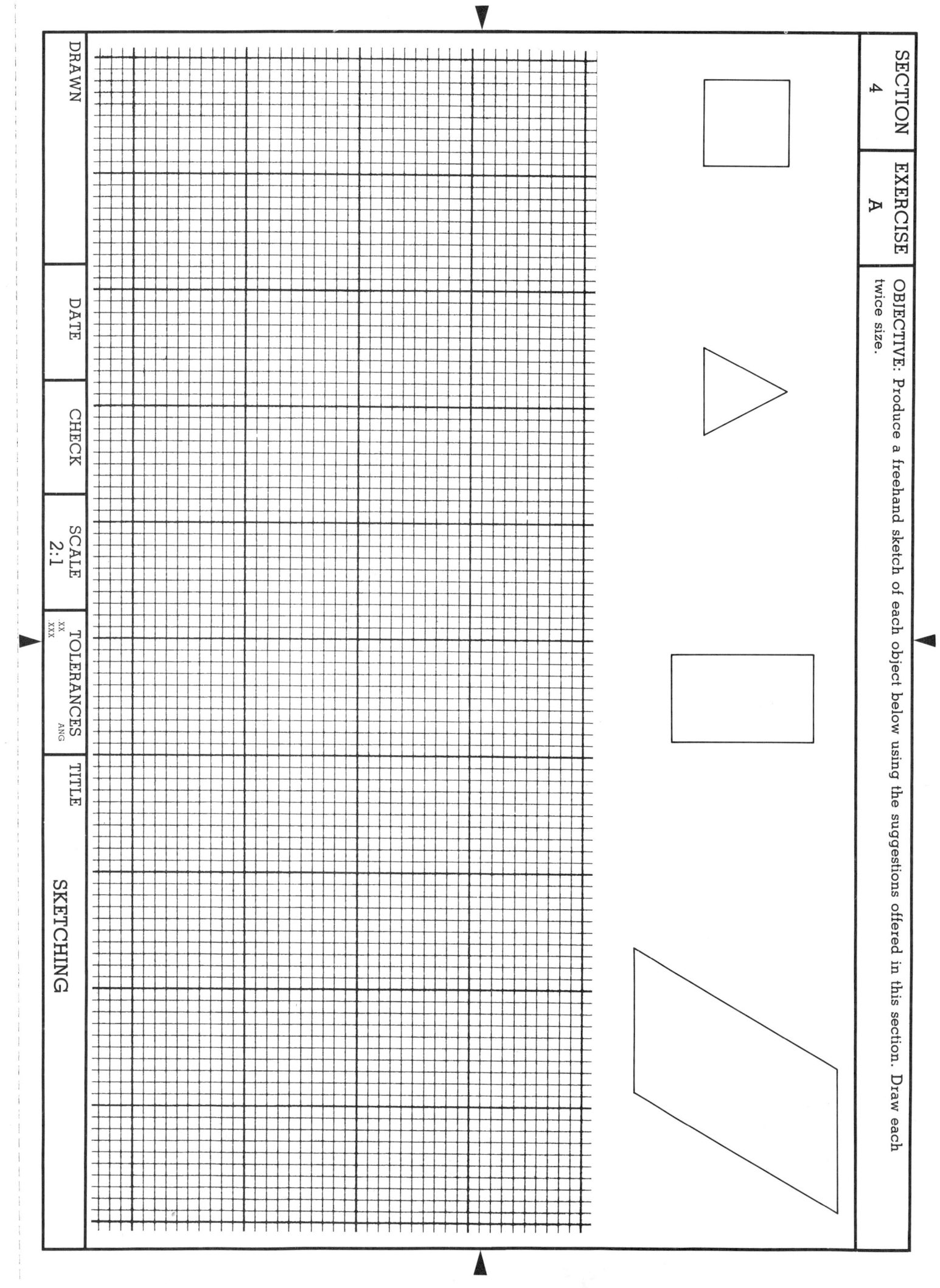

SECTION	EXERCISE	OBJECTIVE: Produce a freehand sketch of each object below using the suggestions offered in this section. Draw each twice size.
4	A	

DRAWN	DATE	CHECK	SCALE 2:1	TOLERANCES XX XXX ANG	TITLE SKETCHING

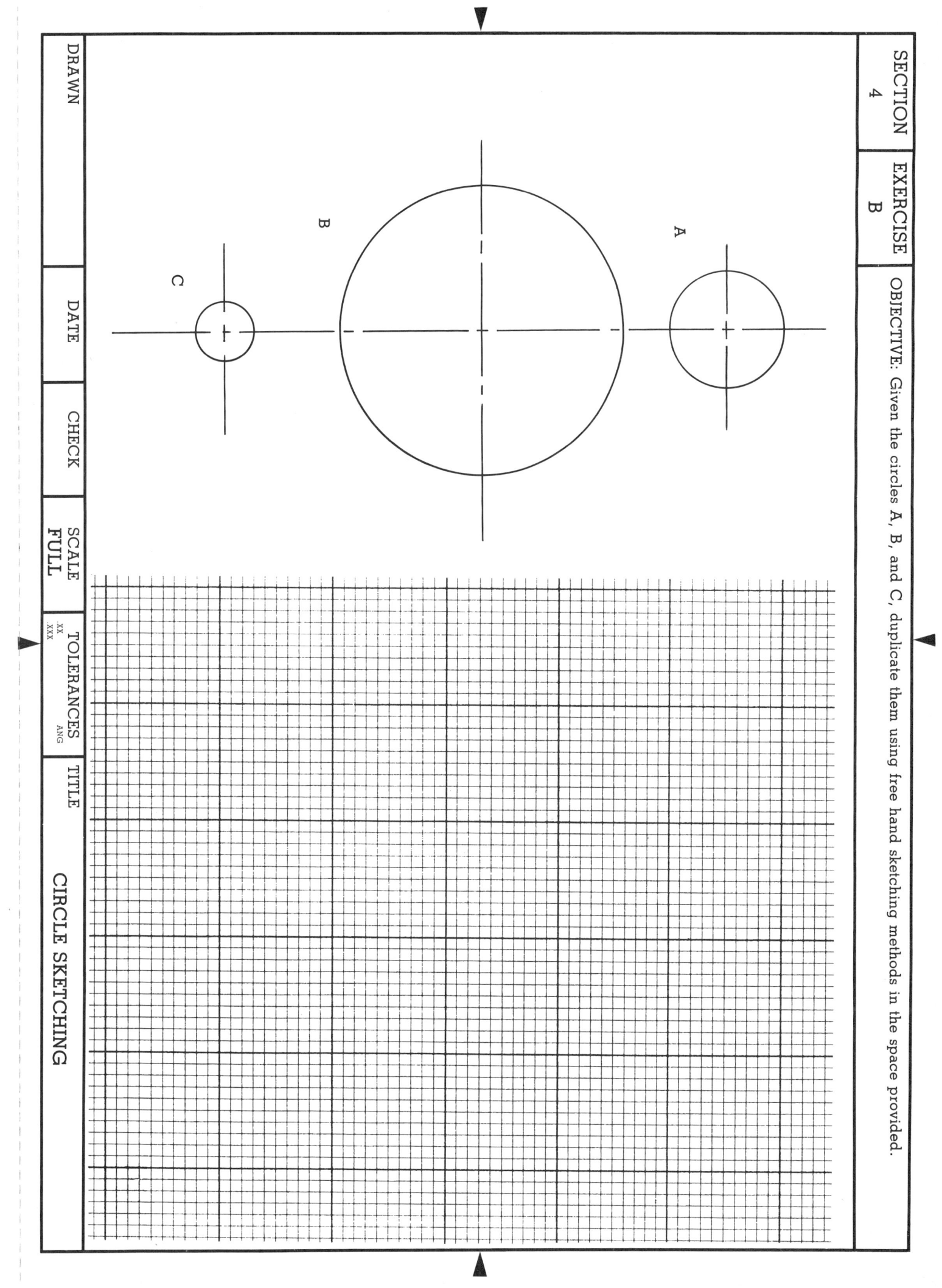

SECTION	EXERCISE	OBJECTIVE: Given the circles A, B, and C, duplicate them using free hand sketching methods in the space provided.
4	B	

DRAWN | **DATE** | **CHECK** | **SCALE** FULL | **TOLERANCES** xx xxx ANG | **TITLE** CIRCLE SKETCHING

SECTION	4
EXERCISE	C

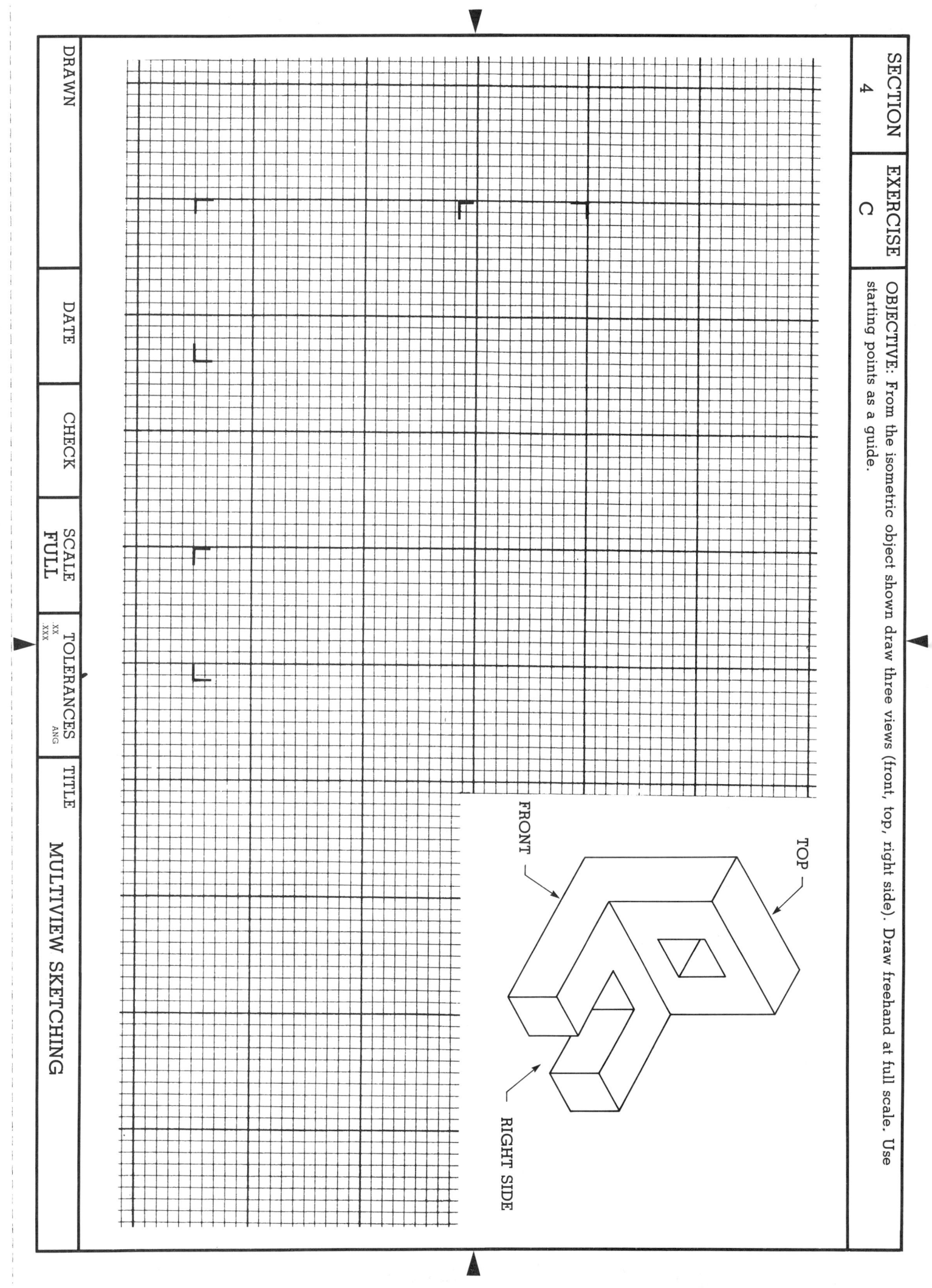

OBJECTIVE: From the isometric object shown draw three views (front, top, right side). Draw freehand at full scale. Use starting points as a guide.

FRONT
TOP
RIGHT SIDE

DRAWN				
DATE	CHECK	SCALE FULL	TOLERANCES XX XXX ANG	

TITLE: MULTIVIEW SKETCHING

SECTION	4
EXERCISE	D

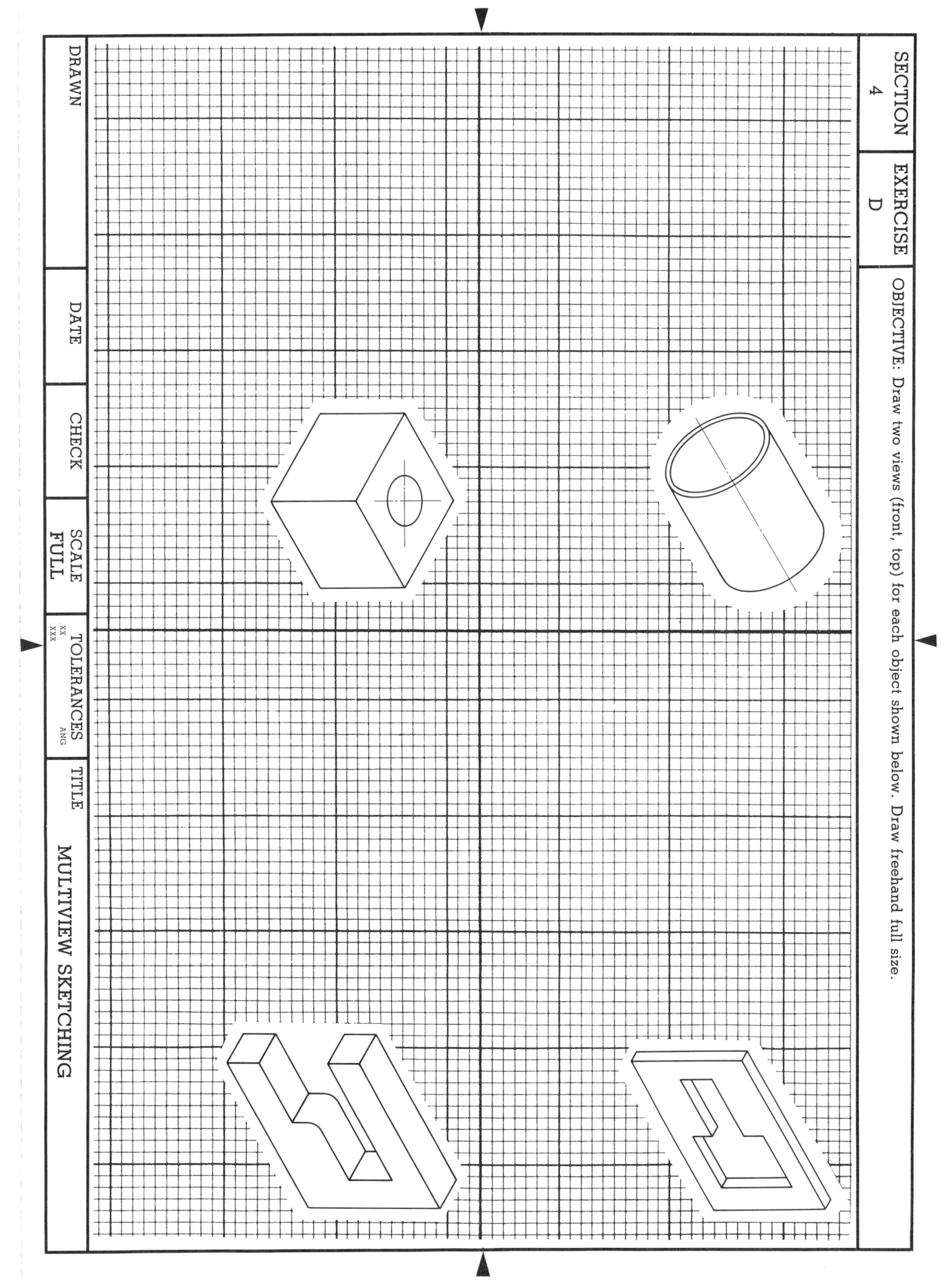

OBJECTIVE: Draw two views (front, top) for each object shown below. Draw freehand full size.

DRAWN	
DATE	
CHECK	
SCALE	FULL
TOLERANCES	XX / XXX / ANG
TITLE	MULTIVIEW SKETCHING

SECTION	EXERCISE			
4	E			

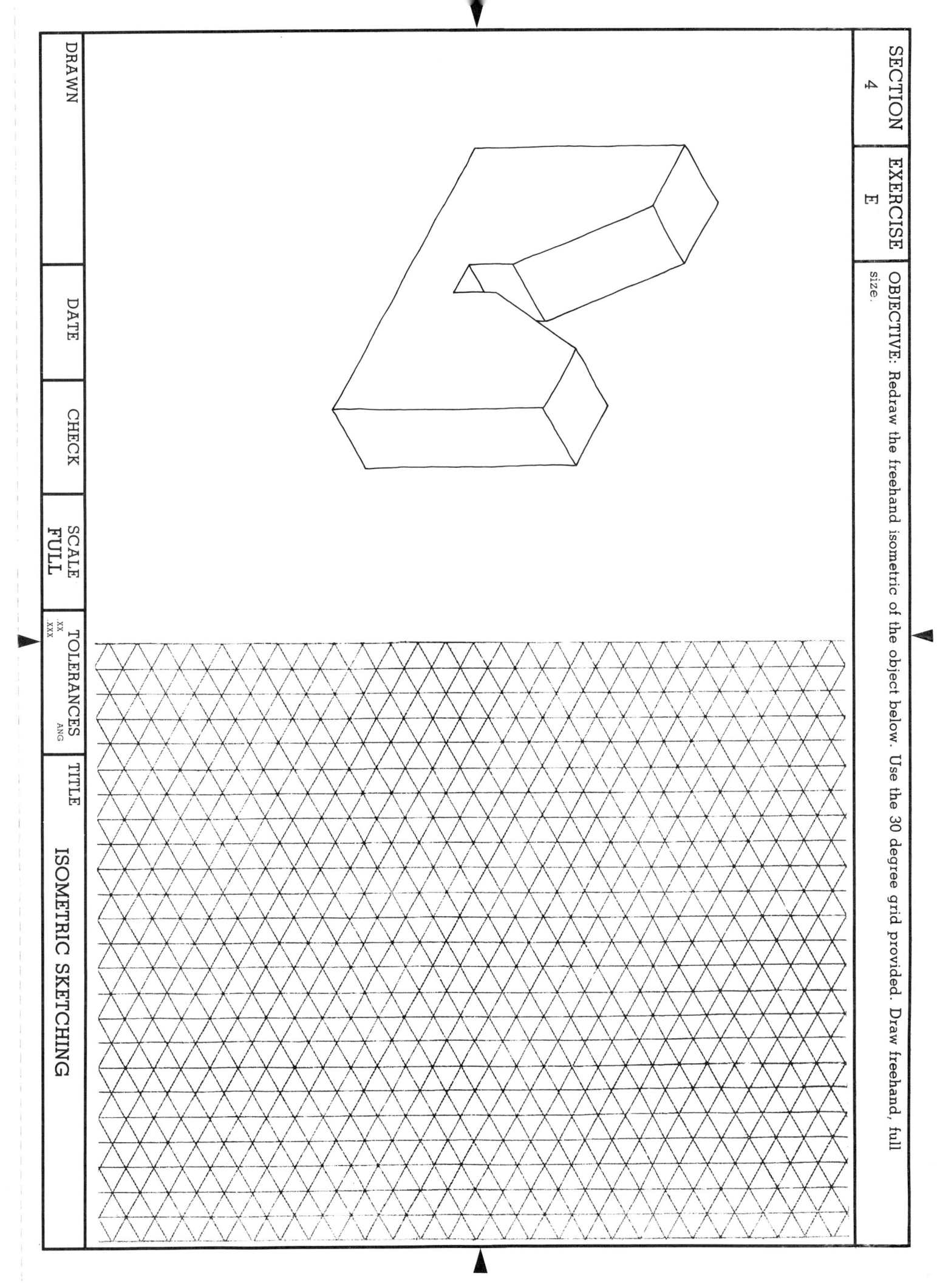

OBJECTIVE: Redraw the freehand isometric of the object below. Use the 30 degree grid provided. Draw freehand, full size.

DRAWN	DATE	CHECK	SCALE FULL	TOLERANCES XX XXX ANG

TITLE: ISOMETRIC SKETCHING

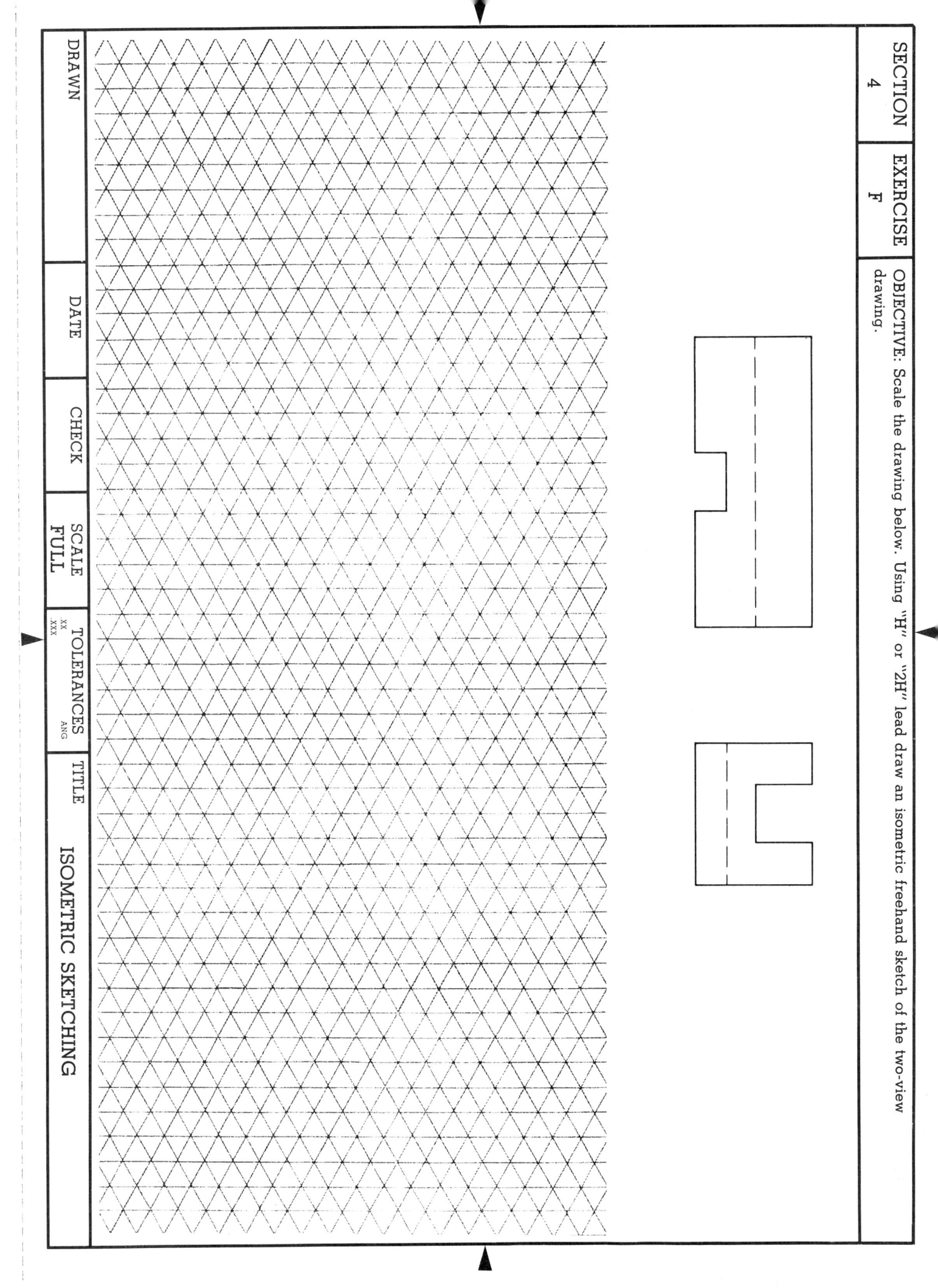

SECTION	EXERCISE	OBJECTIVE: Scale the drawing below. Using "H" or "2H" lead draw an isometric freehand sketch of the two-view drawing.
4	F	

DRAWN	DATE	CHECK	SCALE FULL	TOLERANCES XX XXX ANG	TITLE ISOMETRIC SKETCHING

SECTION	4
EXERCISE	G

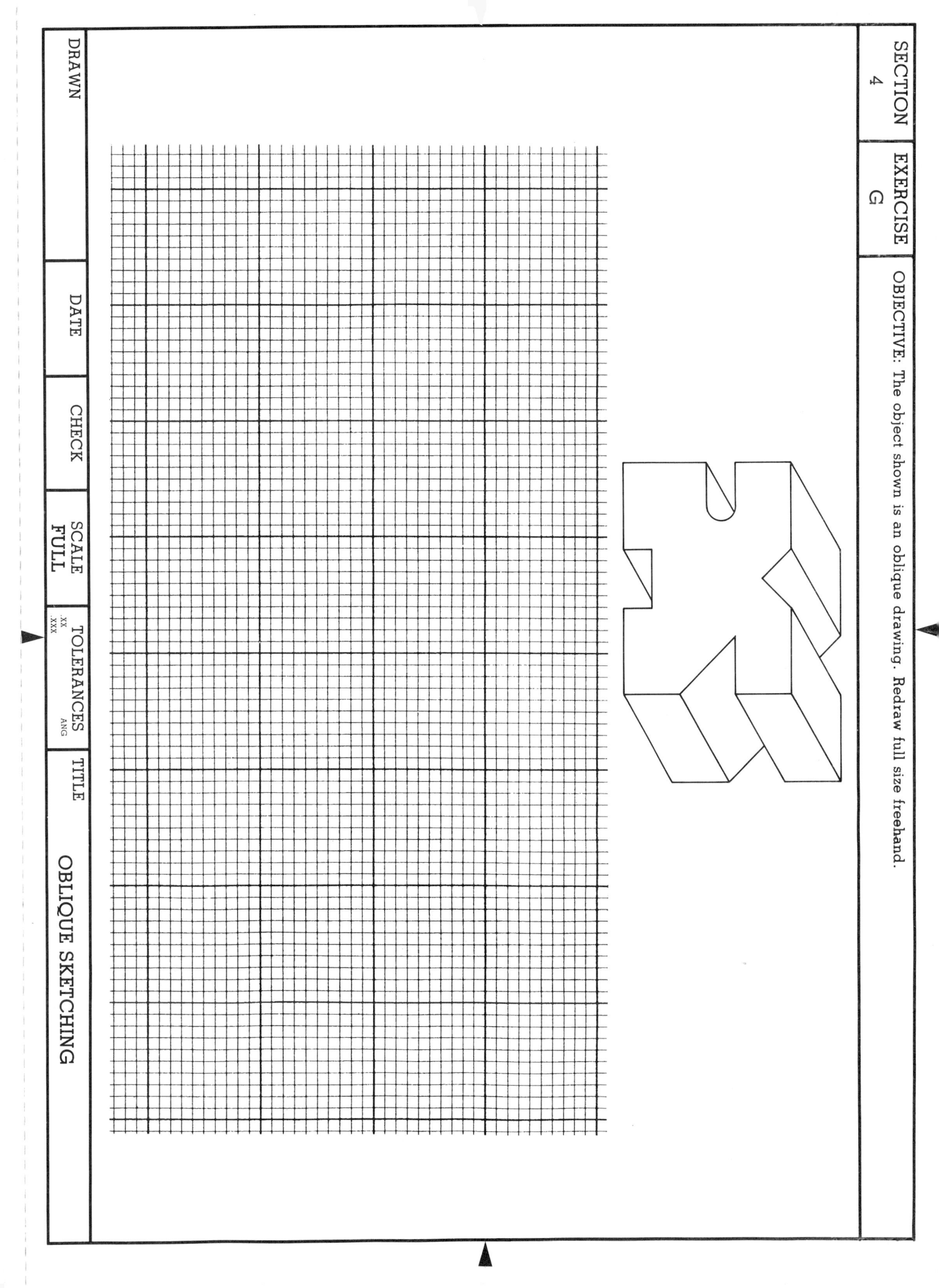

OBJECTIVE: The object shown is an oblique drawing. Redraw full size freehand.

DRAWN	
DATE	
CHECK	
SCALE	FULL
TOLERANCES	XX / XXX / ANG
TITLE	OBLIQUE SKETCHING

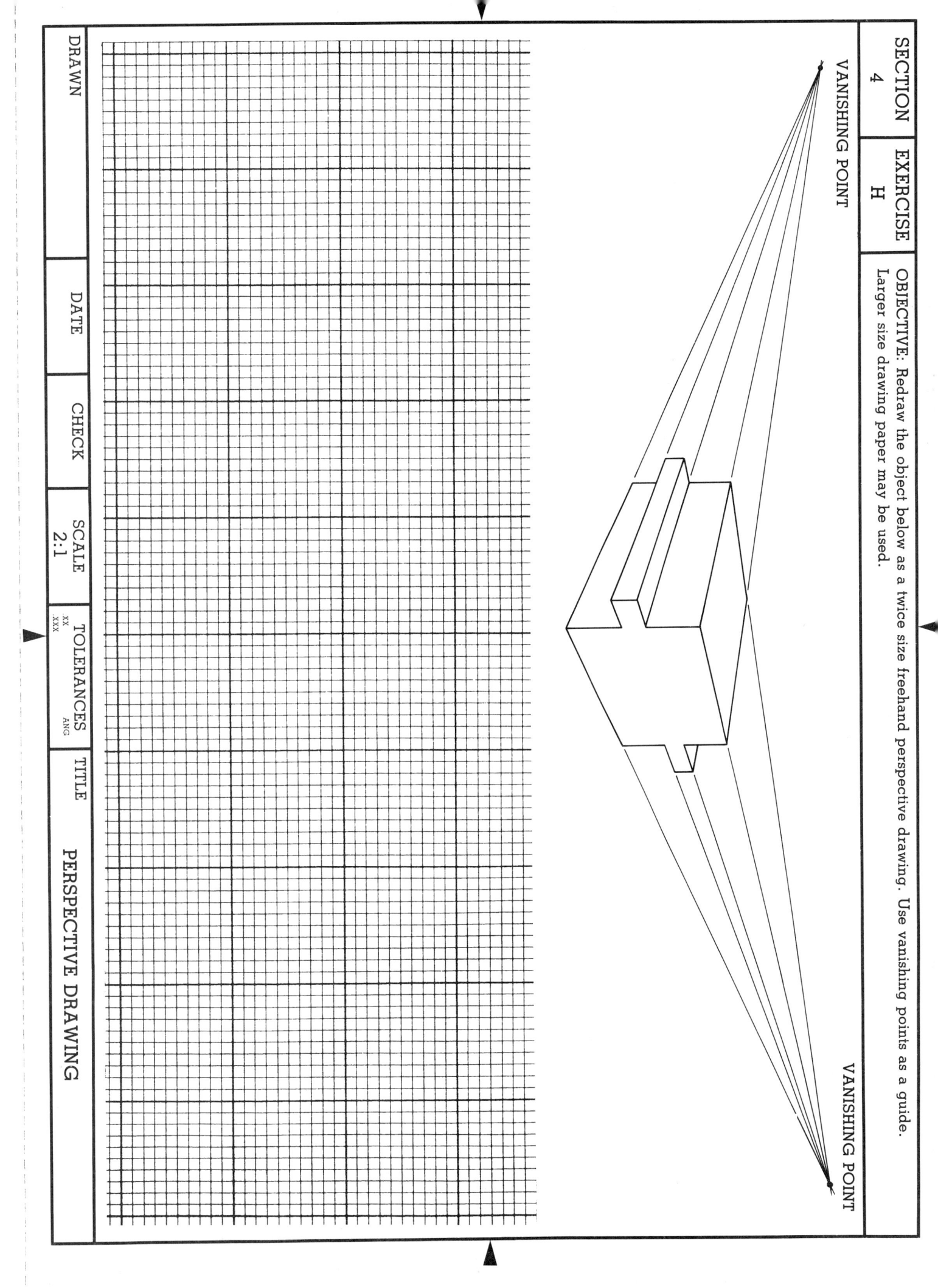

SECTION	EXERCISE	OBJECTIVE: Redraw the object below as a twice size freehand perspective drawing. Use vanishing points as a guide. Larger size drawing paper may be used.
4	H	

VANISHING POINT

VANISHING POINT

DRAWN	DATE	CHECK	SCALE 2:1	TOLERANCES XX XXX ANG	TITLE
					PERSPECTIVE DRAWING

SECTION 5	EXERCISE A

OBJECTIVE: Produce the constructions identified below. Label all points, lines, and intersections. "Heavy up" required lines. Do not erase construction lines.

BISECT A LINE

BISECT AN ACUTE ANGLE

DRAWN	DATE	CHECK	SCALE NONE	TOLERANCES XX XXX ANG	TITLE

GEOMETRIC CONSTRUCTIONS

SECTION	EXERCISE	OBJECTIVE: Produce the constructions identified below. Label all points, lines, and intersections. "Heavy up" required lines. Do not erase construction lines.
5	B	

CONSTRUCT A CIRCLE THROUGH THREE GIVEN POINTS

CONSTRUCT A PARALLEL LINE TO A GIVEN LINE

DRAWN	DATE	CHECK	SCALE NONE	TOLERANCES XX XXX ANG	TITLE GEOMETRIC CONSTRUCTIONS

SECTION	EXERCISE	OBJECTIVE: Produce the constructions identified below. Label all points, lines, and intersections. "Heavy up" required lines. Do not erase construction lines.
5	C	

CONSTRUCT A PERPENDICULAR TO A LINE
FROM A POINT NOT ON THE LINE

DIVIDE A STRAIGHT LINE INTO EQUAL PARTS

DRAWN	DATE	CHECK	SCALE NONE	TOLERANCES XX/XXX ANG	TITLE GEOMETRIC CONSTRUCTIONS

SECTION	EXERCISE	OBJECTIVE: Produce the constructions identified below. Label all points, lines, and intersections. "Heavy up" required lines. Do not erase construction lines.
5	D	

CONSTRUCT AN ANGLE FROM A GIVEN ANGLE

CONSTRUCT A PENGATON WITHIN A CIRCLE

DRAWN	DATE	CHECK	SCALE NONE	TOLERANCES XX XXX ANG	TITLE GEOMETRIC CONSTRUCTIONS

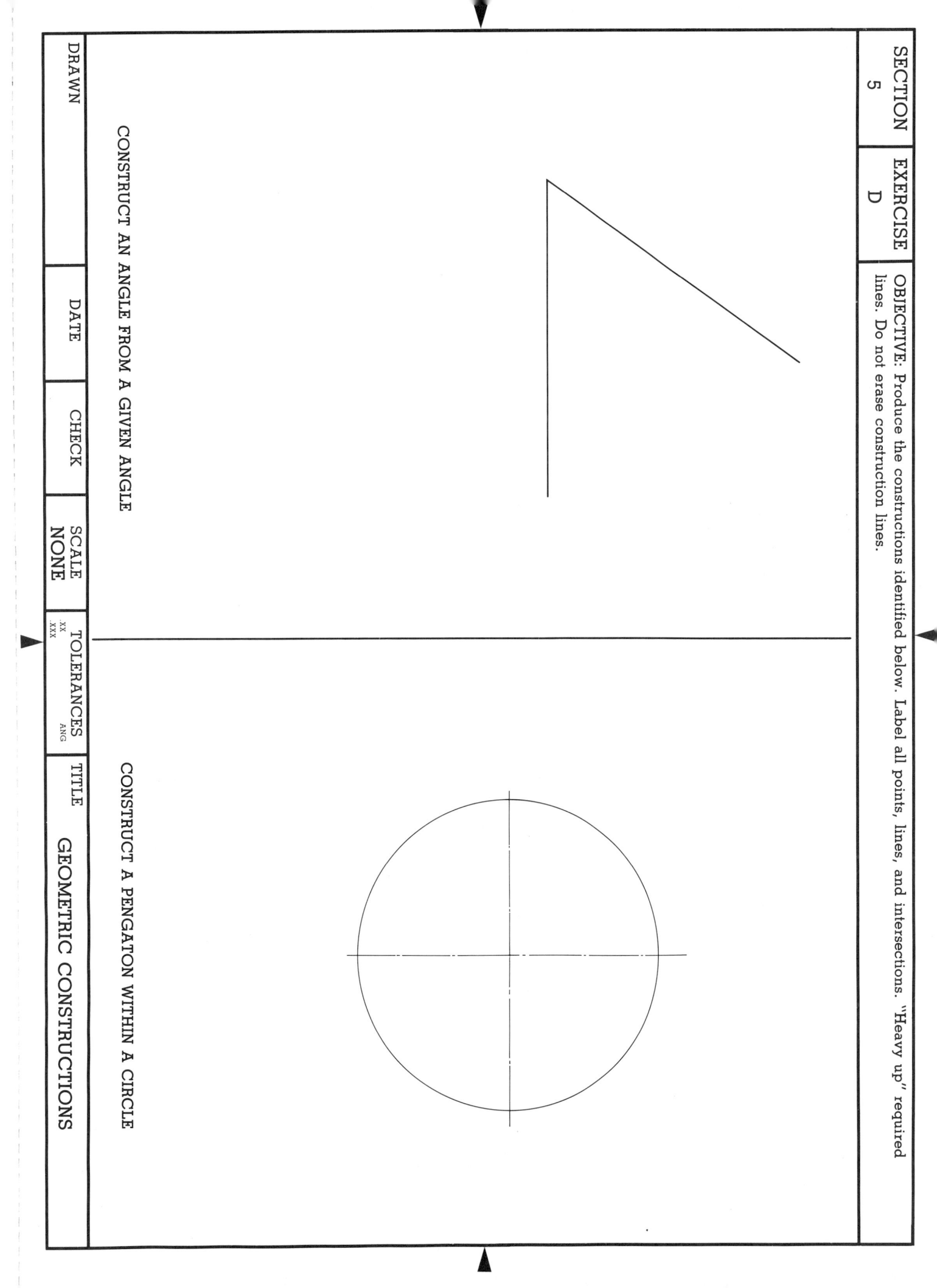

SECTION	EXERCISE	OBJECTIVE: Produce the constructions identified below. Label all points, lines, and intersections. "Heavy up" required lines. Do not erase construction lines.
5	E	

CONSTRUCT A REGULAR HEXAGON WHOSE DISTANCE ACROSS FLATS IS 2.75 INCHES

LOCATE THE CENTER OF A GIVEN CIRCLE

DRAWN	DATE	CHECK	SCALE NONE	TOLERANCES XX XXX ANG	TITLE
					GEOMETRIC CONSTRUCTIONS

SECTION	EXERCISE				
5	F	OBJECTIVE: Produce the constructions identified below. Label all points, lines, and intersections. "Heavy up" required lines. Do not erase construction lines.			

CONSTRUCT TANGENT POINTS (RADIUS) GIVEN AN ACUTE ANGLE AND DISTANCE r

r

CONSTRUCT AN OGEE CURVE GIVEN TWO PARALLEL LINES

DRAWN	DATE	CHECK	SCALE NONE	TOLERANCES XX XXX ANG	TITLE
					GEOMETRIC CONSTRUCTIONS

| SECTION 5 | EXERCISE G | OBJECTIVE: Construct an ellipse using the two-circle method given the information provided below. Label all points and intersections. "Heavy up" required lines. Do not erase construction lines. |

MINOR DIAMETER

MAJOR DIAMETER

| DRAWN | DATE | CHECK | SCALE NONE | TOLERANCES XX XXX ANG | TITLE GEOMETRIC CONSTRUCTION |

| SECTION 5 | EXERCISE H | OBJECTIVE: Construct an ellipse using the two-circle method given the information provided below. Label all points and intersections. "Heavy up" required lines. Do not erase construction lines. |

MINOR DIAMETER

MAJOR DIAMETER

| DRAWN | DATE | CHECK | SCALE NONE | TOLERANCES XX XXX ANG | TITLE GEOMETRIC CONSTRUCTION |

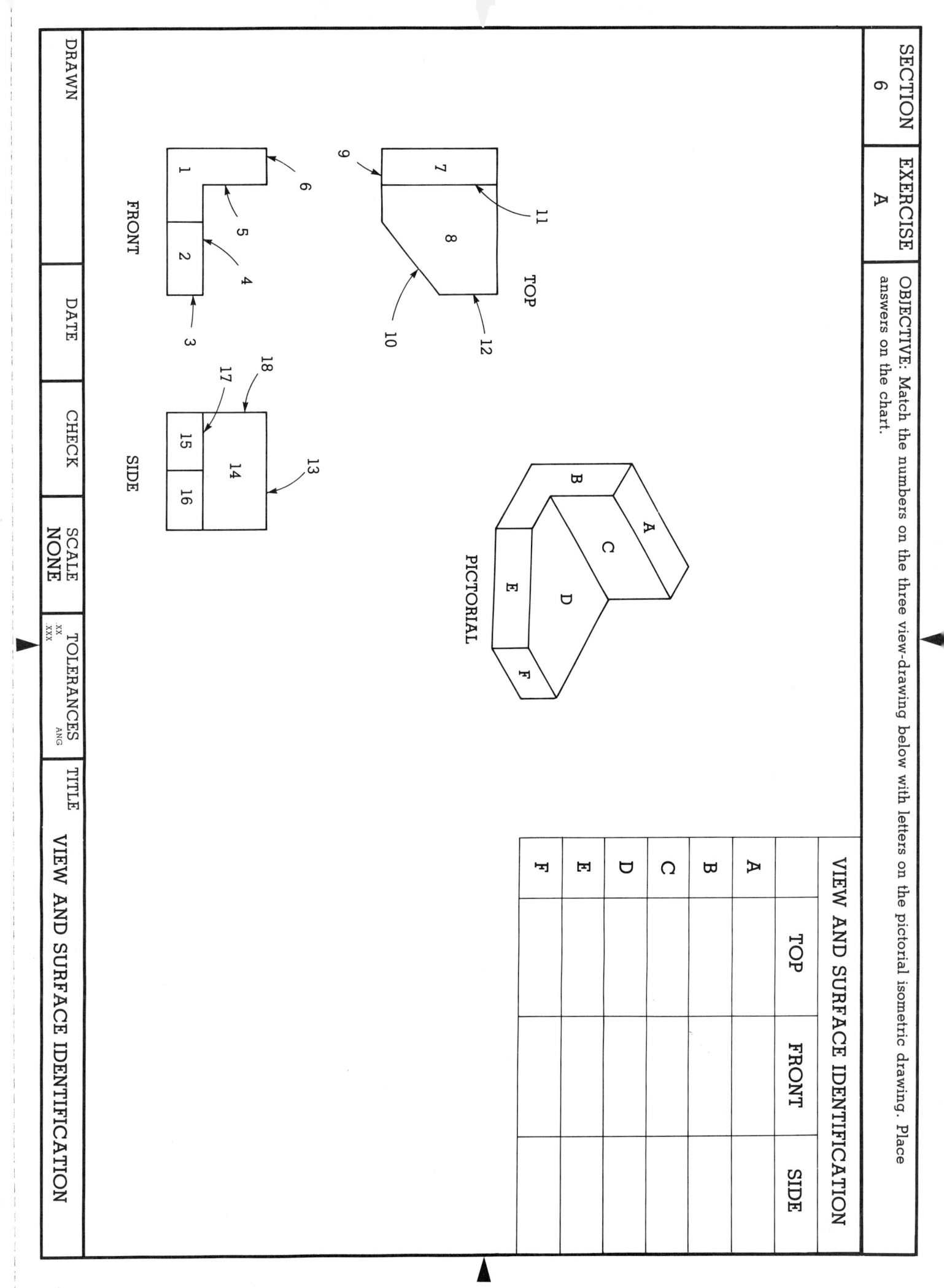

SECTION	EXERCISE		
6	A	OBJECTIVE: Match the numbers on the three view-drawing below with letters on the pictorial isometric drawing. Place answers on the chart.	

TOP

FRONT

SIDE

PICTORIAL

VIEW AND SURFACE IDENTIFICATION

	TOP	FRONT	SIDE
A			
B			
C			
D			
E			
F			

DRAWN	DATE	CHECK	SCALE NONE	TOLERANCES XX XXX ANG	TITLE VIEW AND SURFACE IDENTIFICATION

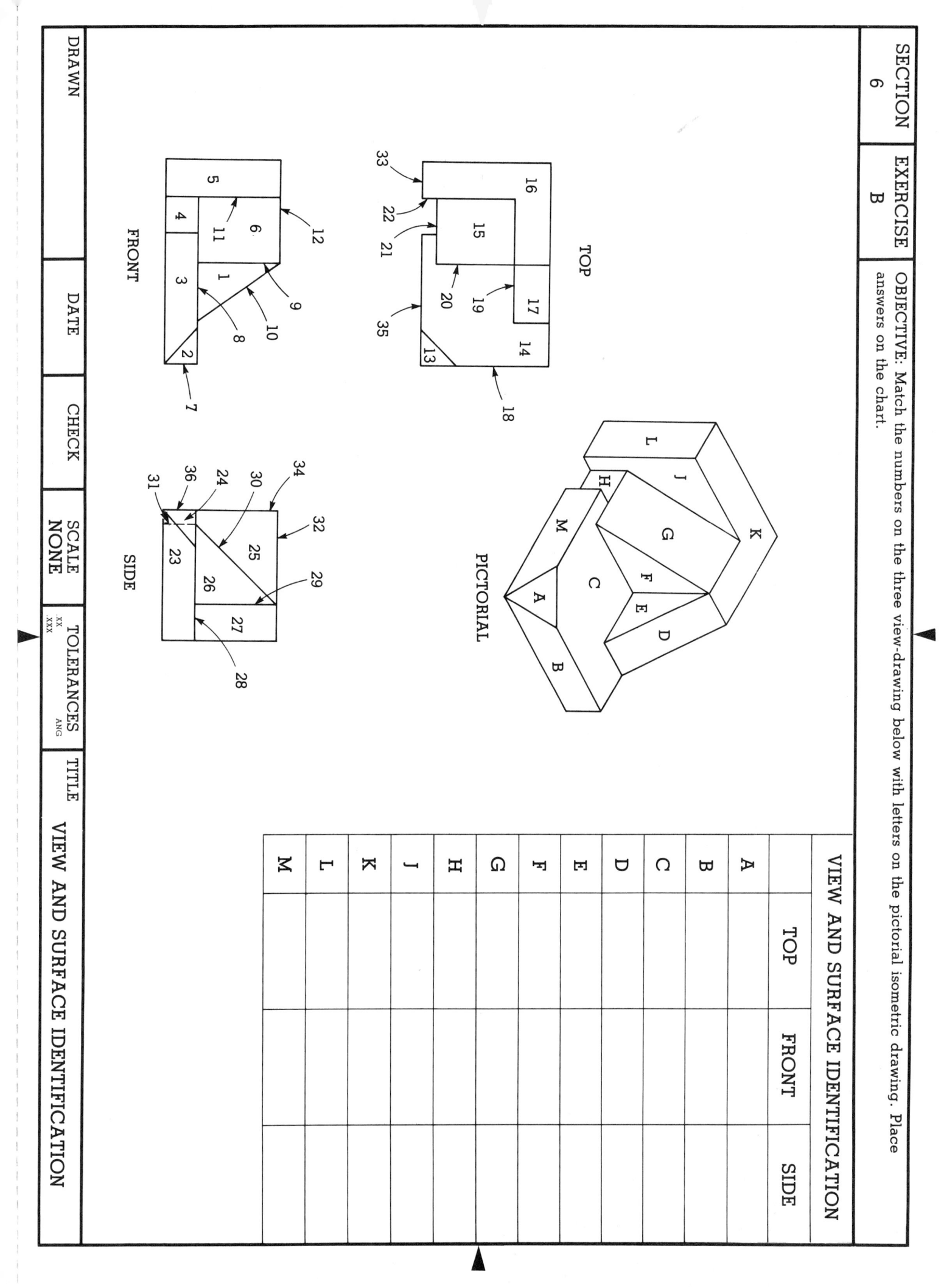

SECTION	EXERCISE	OBJECTIVE: Match the numbers on the three view-drawing below with letters on the pictorial isometric drawing. Place answers on the chart.
6	B	

TOP

FRONT

SIDE

PICTORIAL

VIEW AND SURFACE IDENTIFICATION

	TOP	FRONT	SIDE
A			
B			
C			
D			
E			
F			
G			
H			
J			
K			
L			
M			

DRAWN	DATE	CHECK	SCALE NONE	TOLERANCES XX XXX ANG	TITLE VIEW AND SURFACE IDENTIFICATION

SECTION	EXERCISE
6	C

OBJECTIVE: Produce six different front views for an object whose top view is given. Make all front views the same height.

DRAWN	DATE	CHECK	SCALE NONE	TOLERANCES XX XXX ANG	TITLE **TWO-VIEW DRAWING**

SECTION	EXERCISE	OBJECTIVE: Produce five different front views for an object that has the same given top view. Show front views all the same height.
6	D	

DRAWN	DATE	CHECK	SCALE NONE	TOLERANCES XX XXX ANG	TITLE TWO-VIEW DRAWING

SECTION	EXERCISE
6	E

OBJECTIVE: Draw top and right side views that agree with the given front views of the object shown below. Show all visible, hidden, and center lines. Use orthographic projection principles.

DRAWN	DATE	CHECK	SCALE **NONE**	TOLERANCES XX XXX ANG	TITLE **THREE-VIEW DRAWING**

SECTION 6 EXERCISE F

OBJECTIVE: Add all missing lines to the three-view objects shown below.

TITLE: THREE-VIEW DRAWING
SCALE: NONE

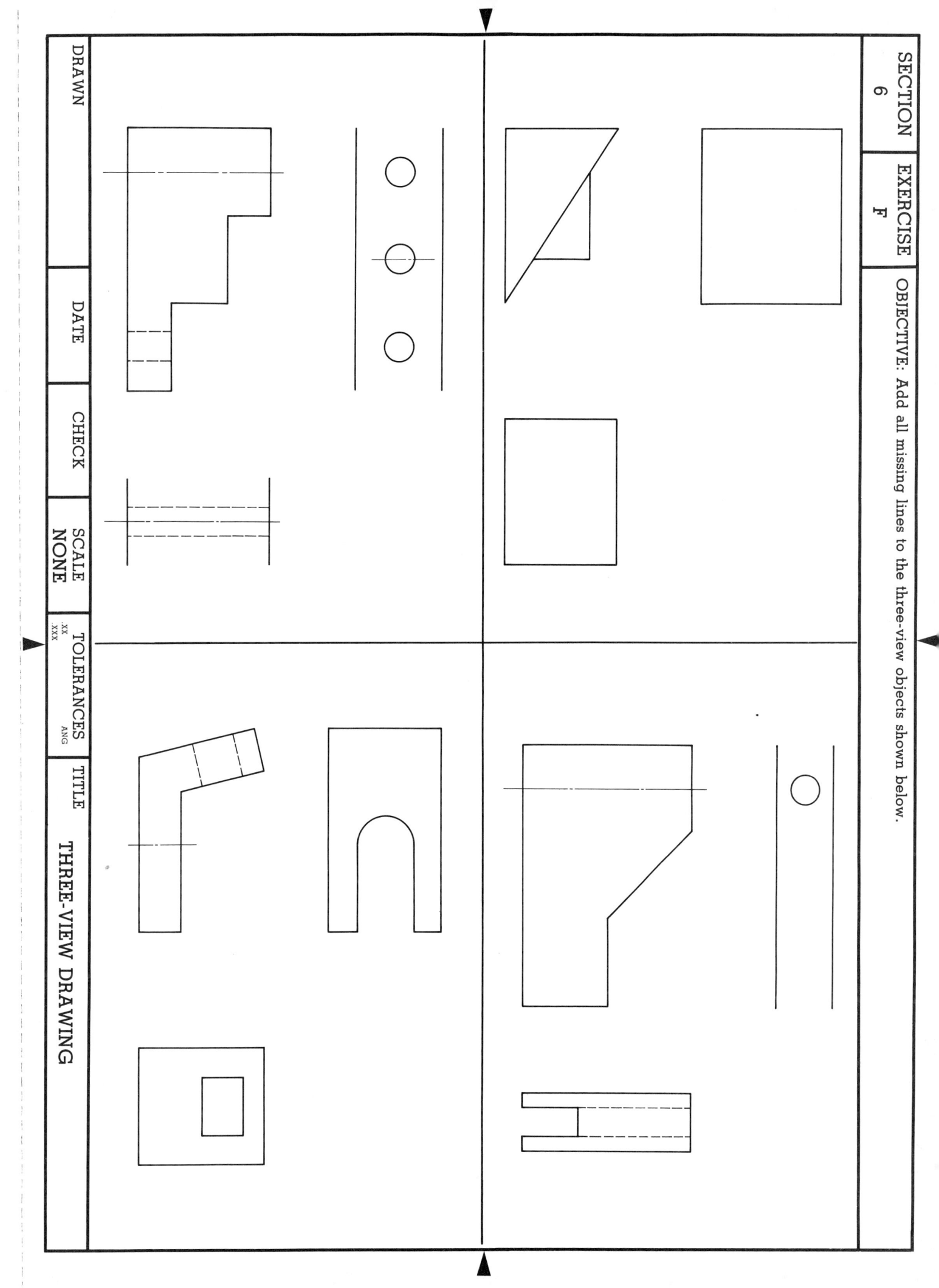

SECTION	EXERCISE	OBJECTIVE: Produce a three-view drawing of the rod bracket shown. Draw full size. No dimensions required.
6	G	

DRAWN	DATE	CHECK	SCALE FULL	TOLERANCES XX XXX ANG	TITLE BRACKET, ROD

SECTION	6
EXERCISE	H

OBJECTIVE: Produce a three-view drawing of the clevis housing shown. Draw full size. No dimensions required.

DRAWN	DATE	CHECK	SCALE FULL	TOLERANCES XX XXX ANG	TITLE **BRACKET, CLEVIS**

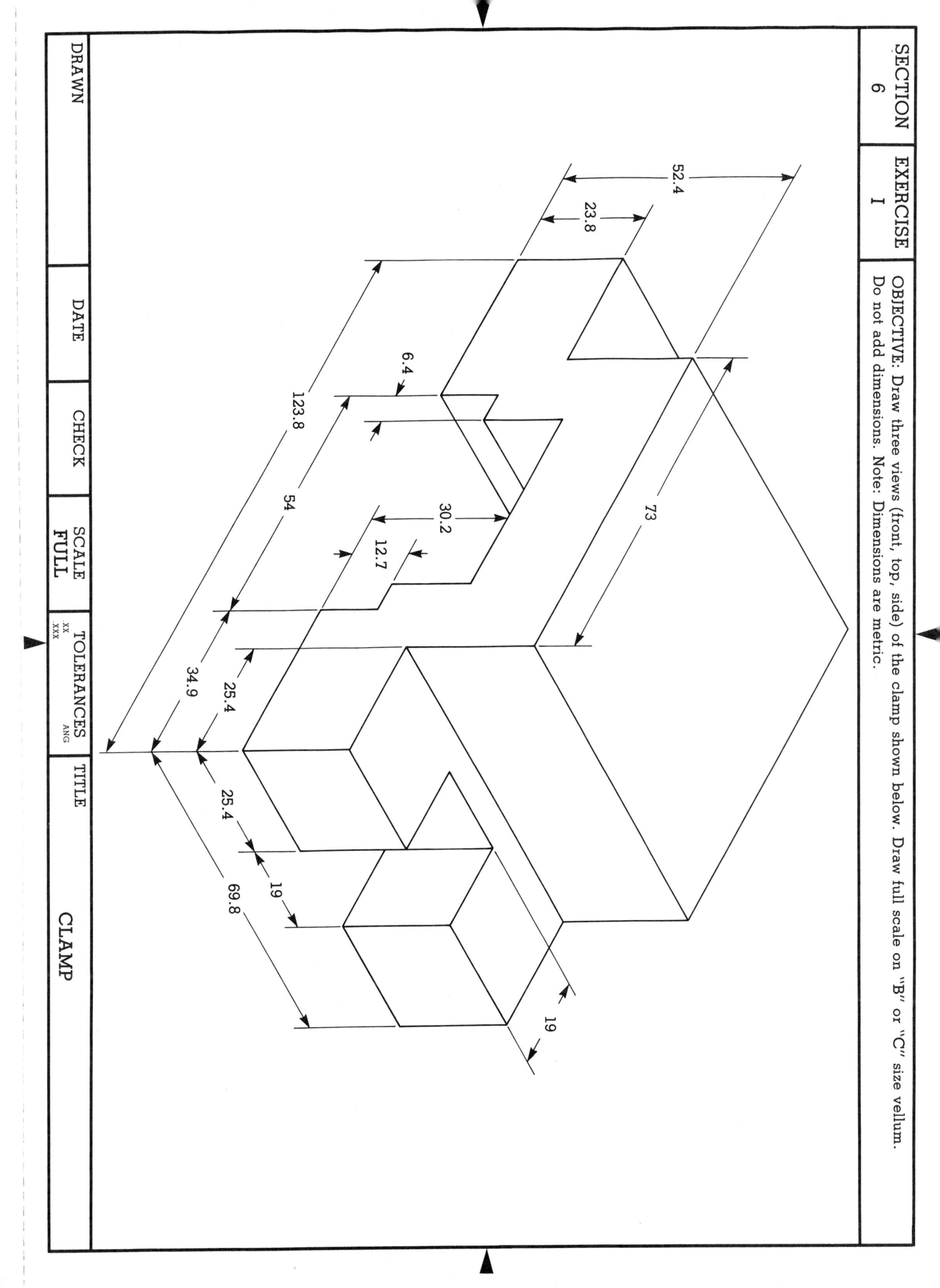

SECTION	EXERCISE	OBJECTIVE: Draw three views (front, top, side) of the clamp shown below. Draw full scale on "B" or "C" size vellum. Do not add dimensions. Note: Dimensions are metric.
6	1	

DRAWN	DATE	CHECK	SCALE FULL	TOLERANCES .XX .XXX ANG	TITLE CLAMP

SECTION	EXERCISE	
6	J	OBJECTIVE: Using the principles of orthographic projection draw six views of the object below. Label each view. Do not add dimensions. Draw full size.

DRAWN	DATE	CHECK	SCALE FULL	TOLERANCES XX / XXX / ANG	TITLE
					SIX-VIEW DRAWING

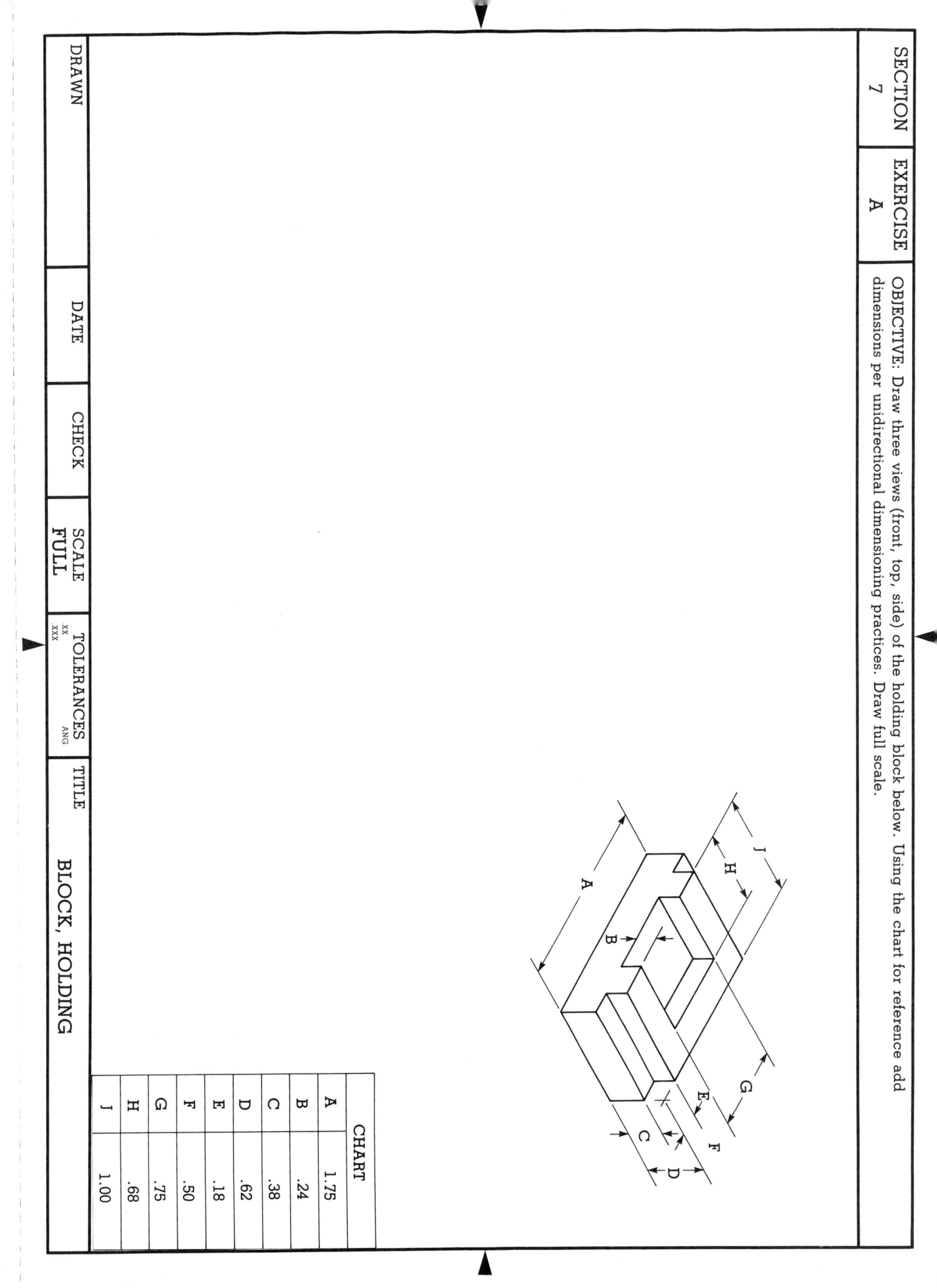

SECTION 7 — EXERCISE A

OBJECTIVE: Draw three views (front, top, side) of the holding block below. Using the chart for reference add dimensions per unidirectional dimensioning practices. Draw full scale.

CHART	
A	1.75
B	.24
C	.38
D	.62
E	.18
F	.50
G	.75
H	.68
J	1.00

TITLE: BLOCK, HOLDING

SCALE: FULL

DRAWN | DATE | CHECK | TOLERANCES (XX, XXX, ANG)

SECTION	EXERCISE	
7	B	OBJECTIVE: Draw three views (front, top, side) of the guide bracket below. Using the chart for reference add metric dimensions per unidirectional dimensioning practices. Draw full scale.

CHART	
A	25.4
B	45.7
C	63.5
D	15.2
E	25.4
F	25.4
G	25.4
H	12.7
J	35.6
K	20.3
L	76.2
M	57.2

DRAWN	DATE	CHECK	SCALE FULL	TOLERANCES XX XXX ANG	TITLE BRACKET, GUIDE

SECTION	EXERCISE	OBJECTIVE: Convert the fractional dimensions below to two-place decimal-inch and millimeter dimensions. Use good lettering practices.
7	C	

FRACTIONAL	DECIMAL INCH	METRIC (MILLIMETERS)
4 3/8		
1 7/16		
3 1/2		
2 13/16		
3/4		
1/32		
7 11/32		
6 1/4		
1 33/64		
5 5/16		
3/64		
1/8		
2 5/8		
4 29/64		
8 3/32		
11/16		
3 5/32		

DRAWN	DATE	CHECK	SCALE NONE	TOLERANCES XX XXX ANG	TITLE DIMENSIONAL CONVERSION

SECTION	EXERCISE
7	D

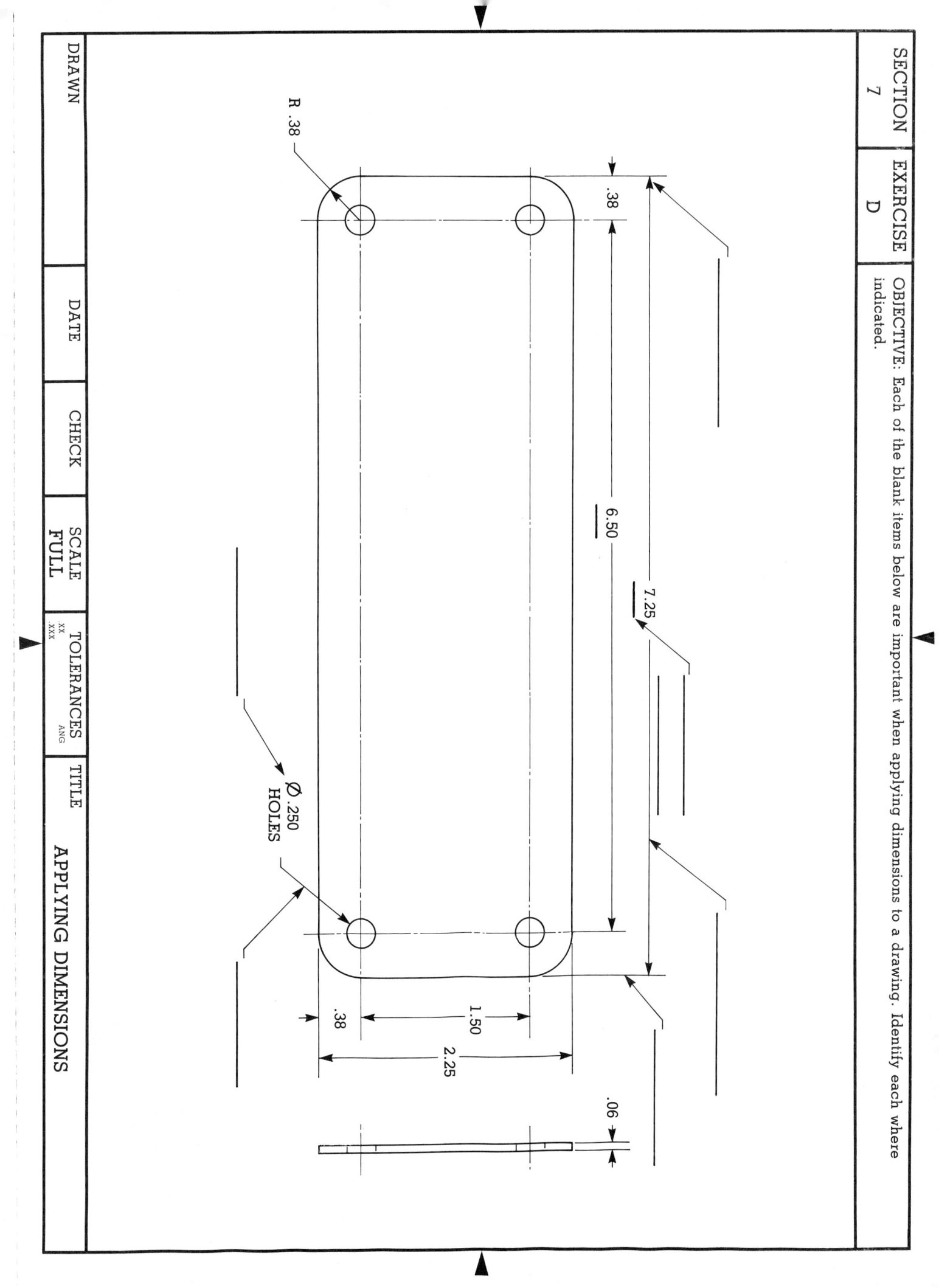

OBJECTIVE: Each of the blank items below are important when applying dimensions to a drawing. Identify each where indicated.

R .38

.38

6.50

7.25

Ø .250
HOLES

.38

1.50

2.25

.06

DRAWN	DATE	CHECK	SCALE FULL	TOLERANCES XX XXX ANG	TITLE

APPLYING DIMENSIONS

SECTION	EXERCISE
7	E

OBJECTIVE: Dimension only the various diameters on the pulley below per the information provided. Use acceptable dimensioning practices. Identify all dimensions in both decimal inch and metric units.

KEY

OUTSIDE RIM DIAMETER = 3.50

INSIDE RIM DIAMETER = 3.00

SHAFT HOLE DIAMETER = .50

RELIEF HOLE DIAMETERS (3) = .62

CENTER DISTANCE FOR RELIEF HOLES = 2.06

DRAWN	DATE	CHECK	SCALE FULL	TOLERANCES XX XXX ANG	TITLE DIMENSIONING DIAMETERS

SECTION	EXERCISE	\multicolumn{4}{l	}{OBJECTIVE: Scale each figure below and apply a decimal inch radius dimension to each. Be accurate within .03 inch. Use acceptable drafting practices.}
7	F				

DRAWN	DATE	CHECK	SCALE **FULL**	TOLERANCES xx xxx ANG	TITLE **APPLYING DIMENSIONS**

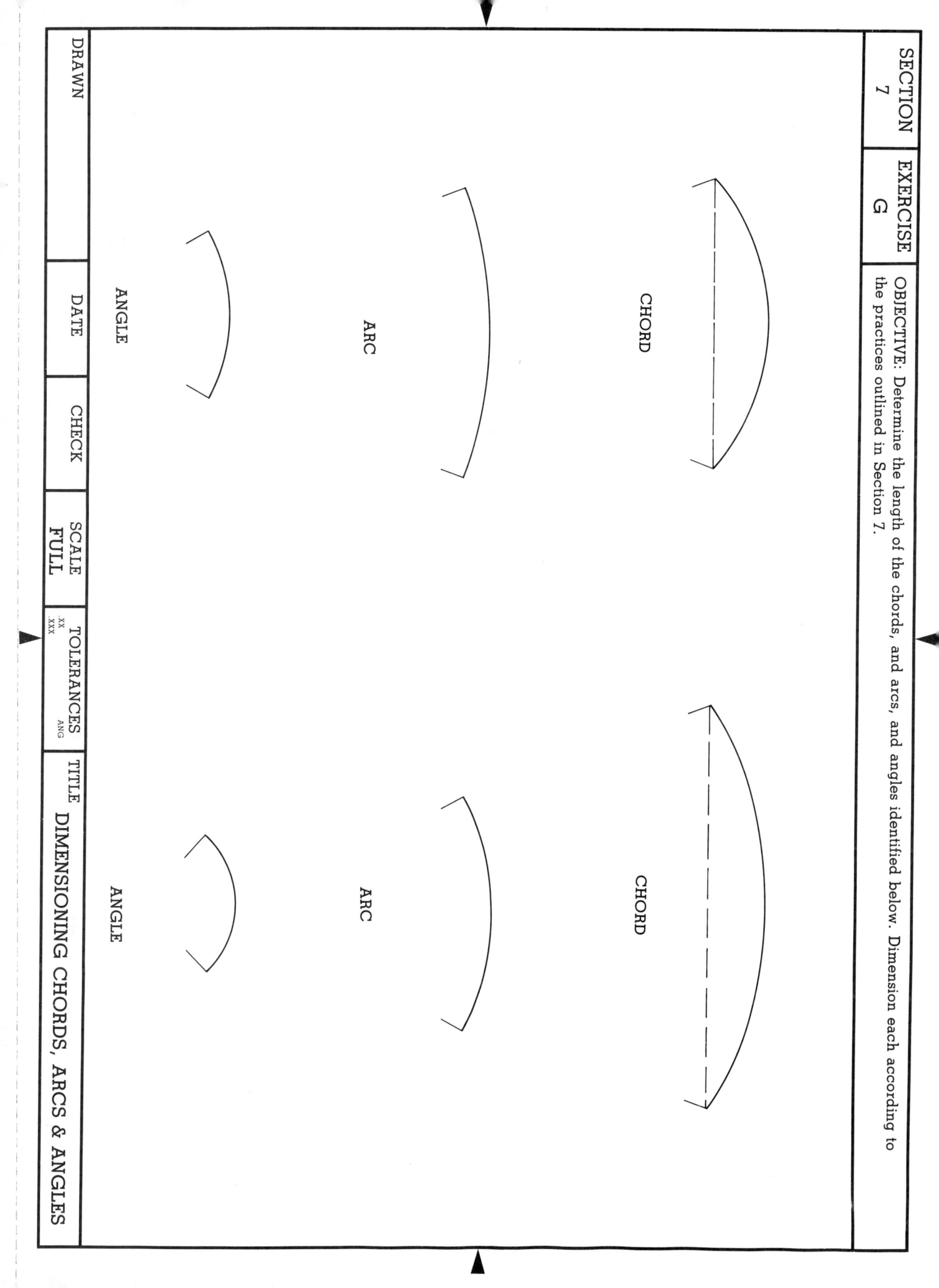

| SECTION 7 | EXERCISE G | OBJECTIVE: Determine the length of the chords, and arcs, and angles identified below. Dimension each according to the practices outlined in Section 7. |

CHORD

ARC

ANGLE

CHORD

ARC

ANGLE

| DRAWN | DATE | CHECK | SCALE FULL | TOLERANCES XX XXX ANG | TITLE DIMENSIONING CHORDS, ARCS & ANGLES |

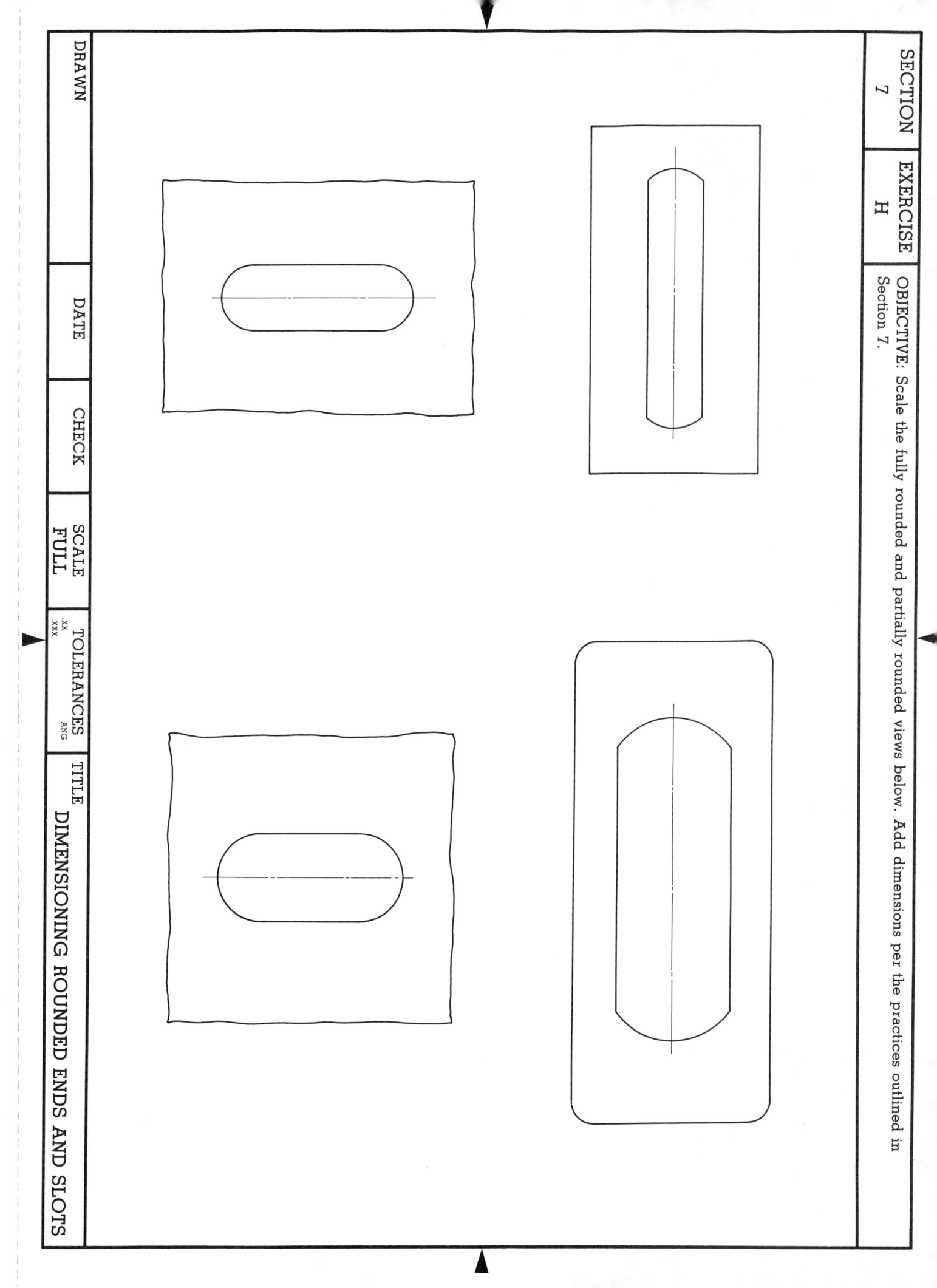

SECTION	EXERCISE	OBJECTIVE: Scale the fully rounded and partially rounded views below. Add dimensions per the practices outlined in Section 7.
7	H	

DRAWN	DATE	CHECK	SCALE FULL	TOLERANCES XX XXX ANG	TITLE DIMENSIONING ROUNDED ENDS AND SLOTS

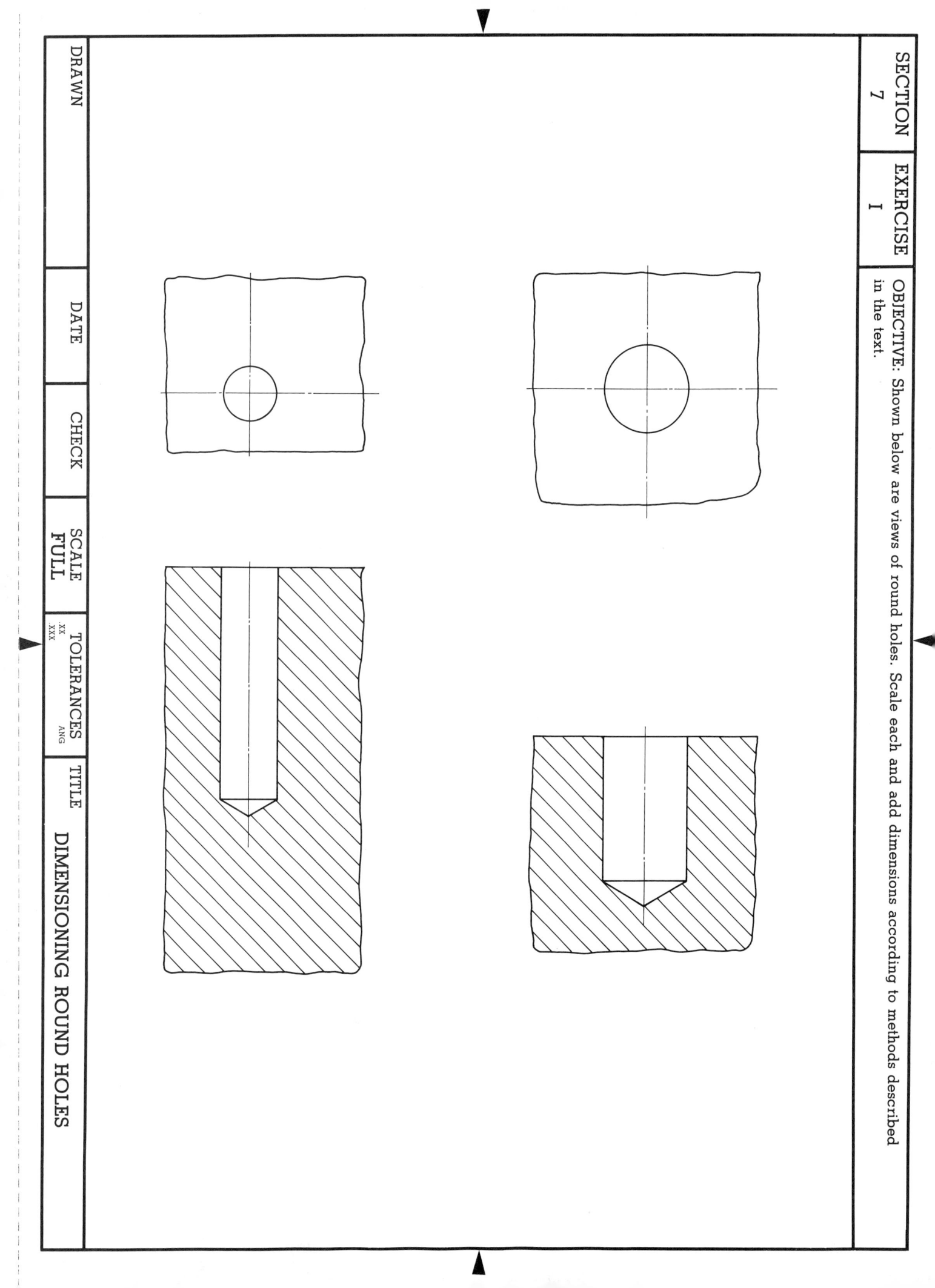

SECTION	EXERCISE	OBJECTIVE: Shown below are views of round holes. Scale each and add dimensions according to methods described in the text.
7	1	

DRAWN	DATE	CHECK	SCALE FULL	TOLERANCES XX XXX ANG	TITLE
					DIMENSIONING ROUND HOLES

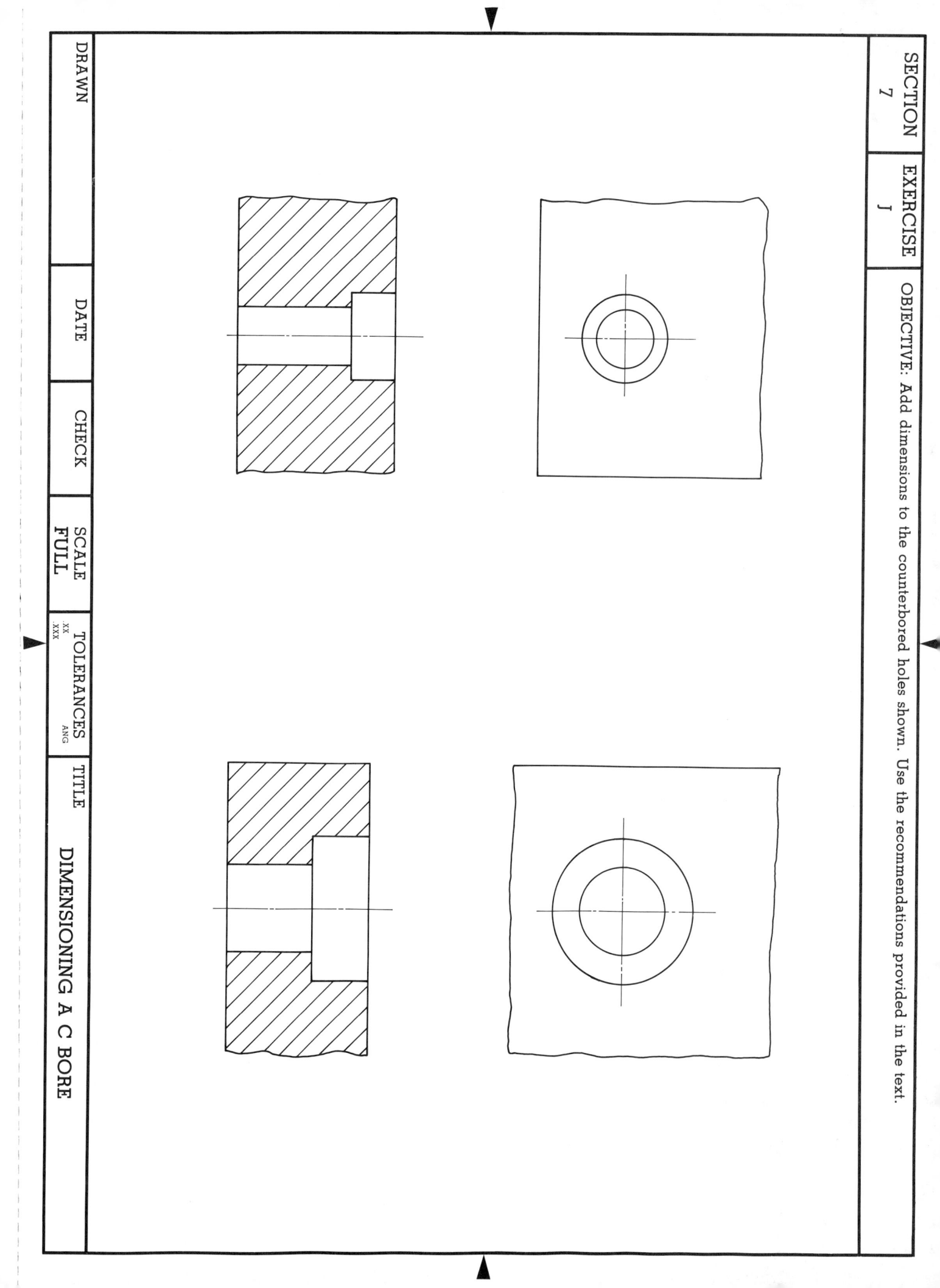

SECTION	EXERCISE	OBJECTIVE: Add dimensions to the counterbored holes shown. Use the recommendations provided in the text.
7	J	

DRAWN	DATE	CHECK	SCALE FULL	TOLERANCES XX / XXX / ANG	TITLE DIMENSIONING A C BORE

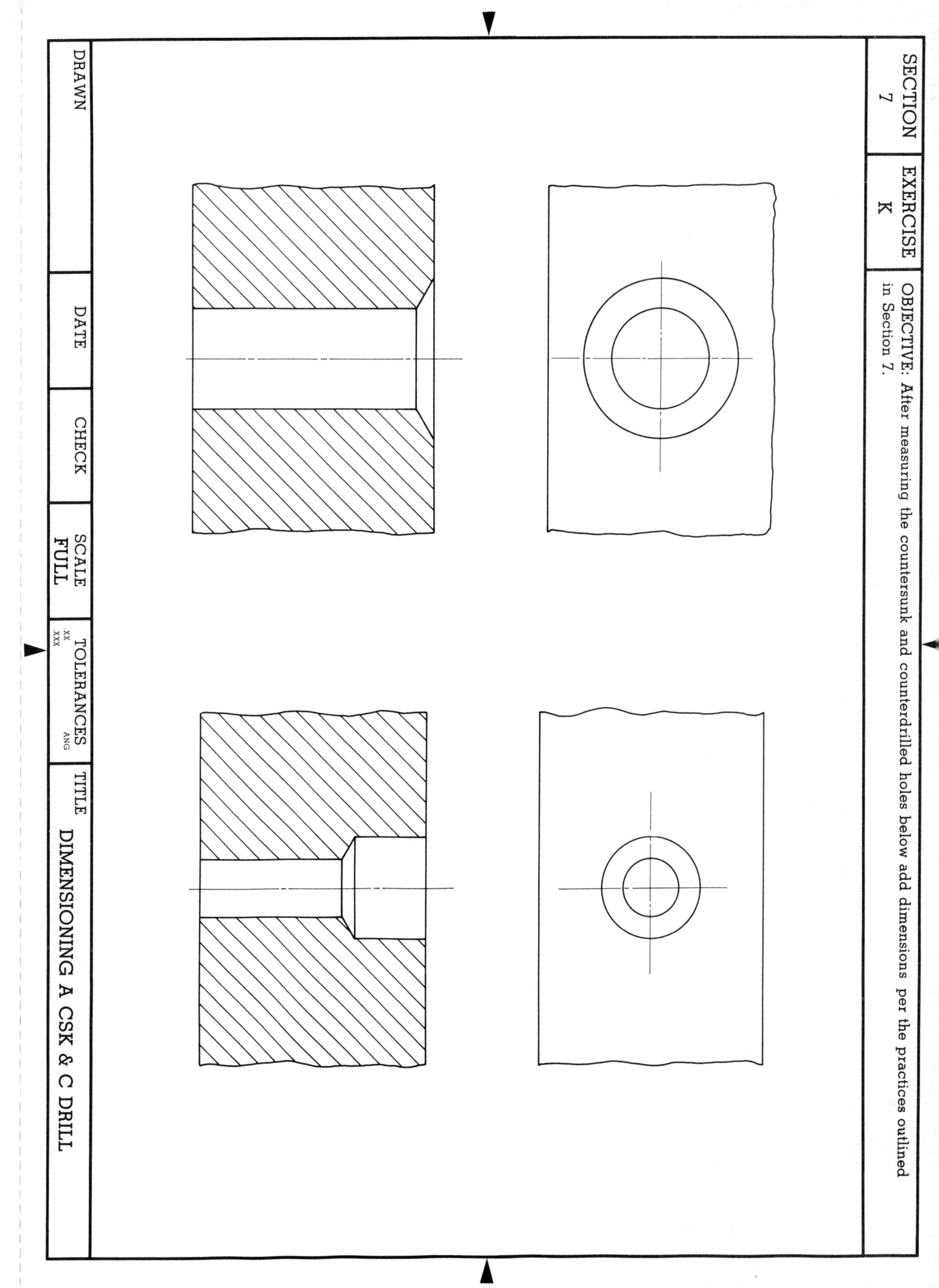

SECTION	7
EXERCISE	K

OBJECTIVE: After measuring the countersunk and counterdrilled holes below add dimensions per the practices outlined in Section 7.

DRAWN	
DATE	
CHECK	
SCALE	FULL
TOLERANCES	.XX .XXX ANG
TITLE	DIMENSIONING A CSK & C DRILL

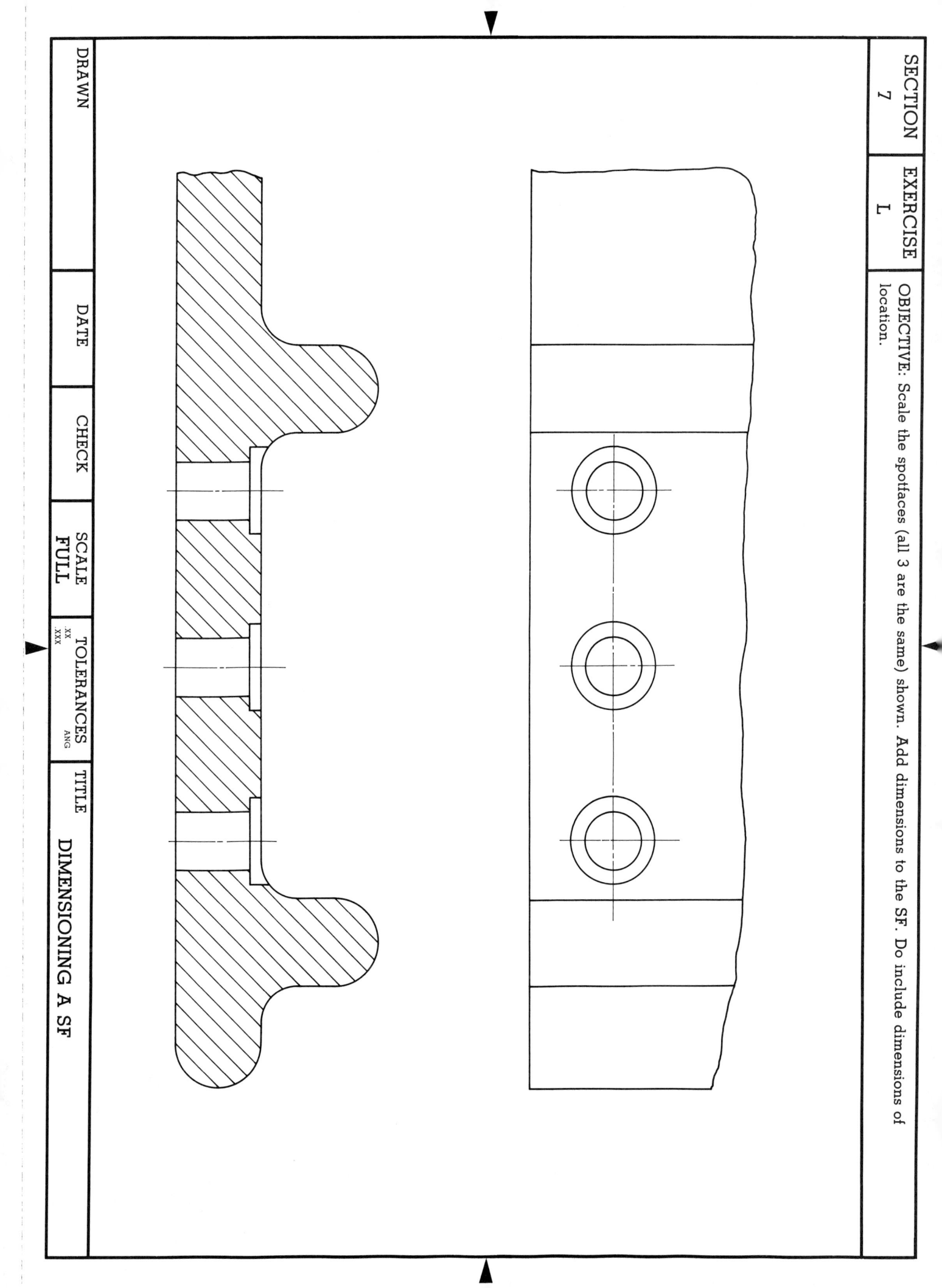

SECTION	EXERCISE	
7	L	OBJECTIVE: Scale the spotfaces (all 3 are the same) shown. Add dimensions to the SF. Do include dimensions of location.

DRAWN	DATE	CHECK	SCALE FULL	TOLERANCES XX XXX ANG	TITLE
					DIMENSIONING A SF

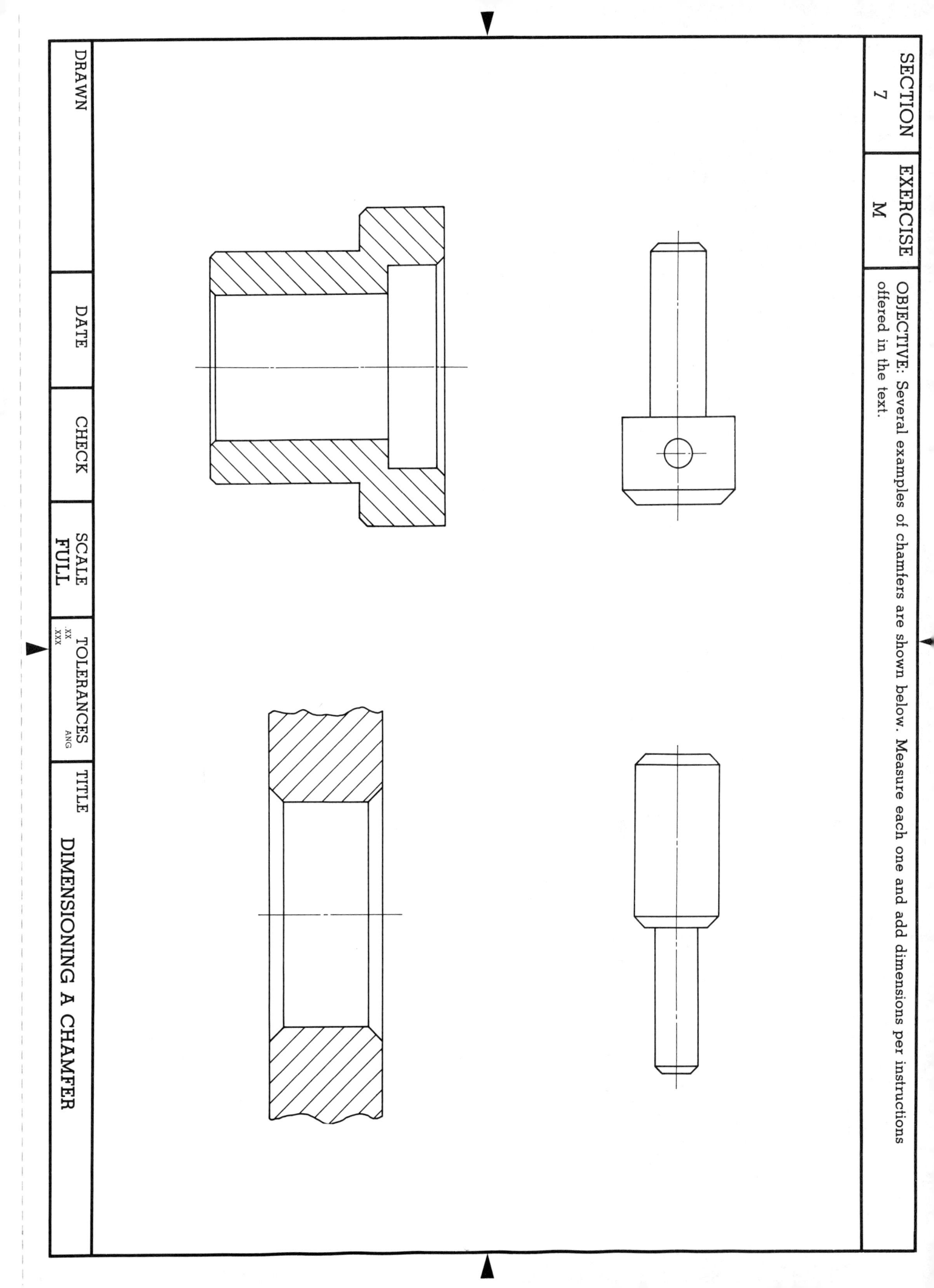

SECTION	7
EXERCISE	M

OBJECTIVE: Several examples of chamfers are shown below. Measure each one and add dimensions per instructions offered in the text.

DRAWN	
DATE	
CHECK	
SCALE	FULL
TOLERANCES	XX / XXX / ANG
TITLE	DIMENSIONING A CHAMFER

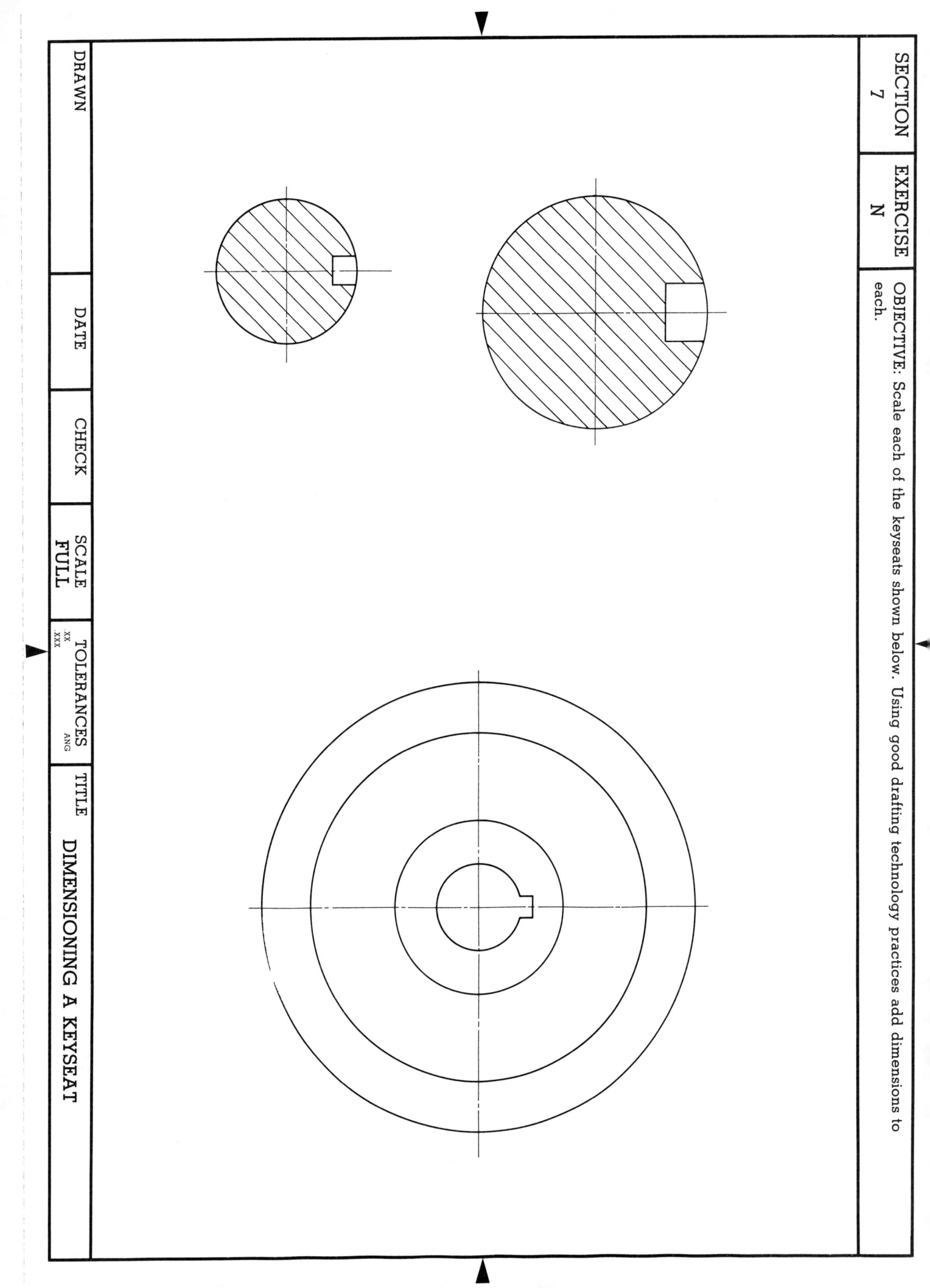

SECTION	7
EXERCISE	N

OBJECTIVE: Scale each of the keyseats shown below. Using good drafting technology practices add dimensions to each.

DRAWN	
DATE	
CHECK	
SCALE	FULL
TOLERANCES	XX / XXX / ANG
TITLE	DIMENSIONING A KEYSEAT

SECTION 7

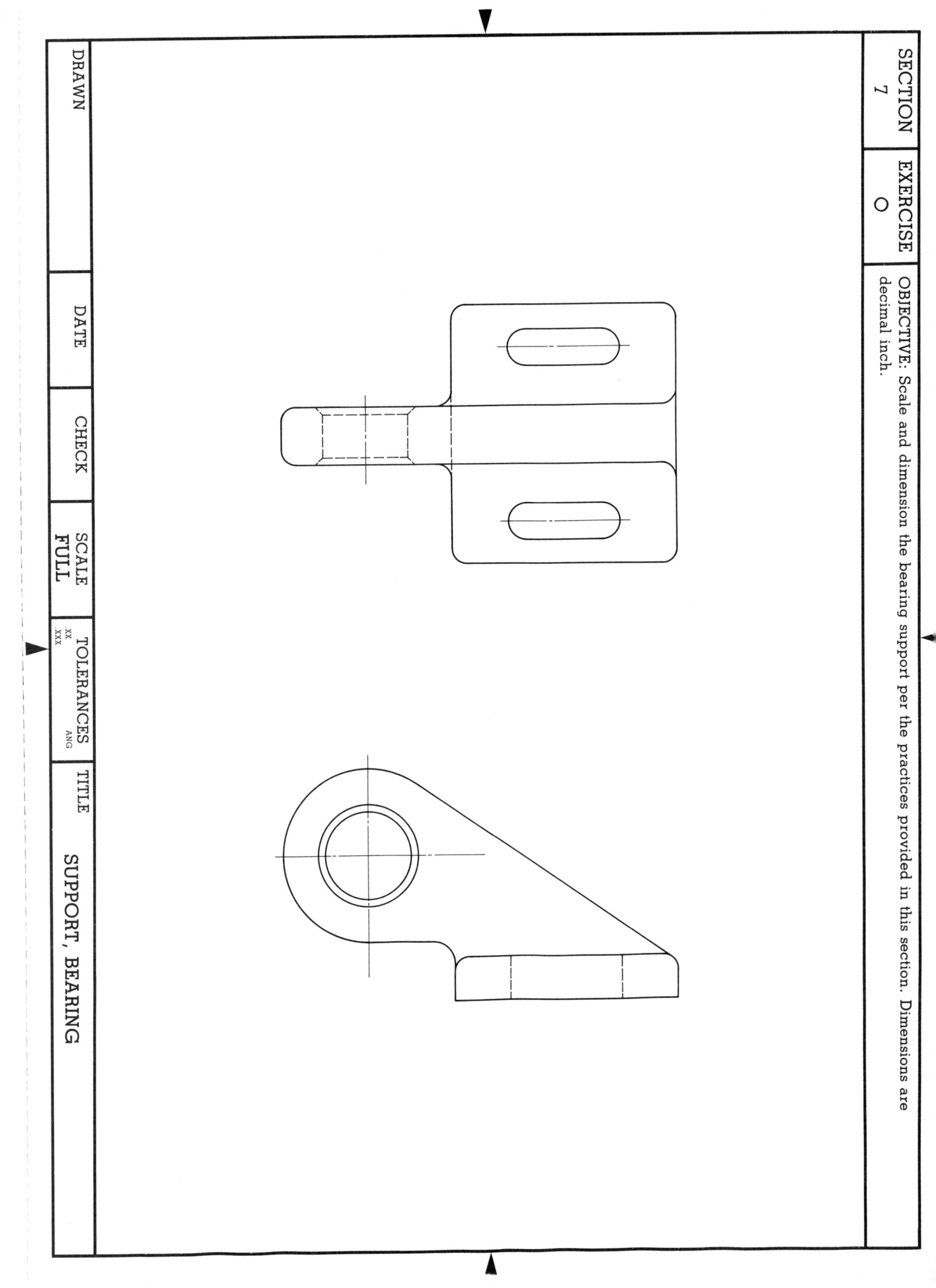

EXERCISE 0

OBJECTIVE: Scale and dimension the bearing support per the practices provided in this section. Dimensions are decimal inch.

| DRAWN | DATE | CHECK | SCALE FULL | TOLERANCES XX XXX ANG | TITLE SUPPORT, BEARING |

SECTION	EXERCISE	OBJECTIVE: Scale and dimension the mounting plate according to acceptable drafting practices. Dimensions are metric.
7	P	

DRAWN	DATE	CHECK	SCALE FULL	TOLERANCES xx xxx ANG	TITLE PLATE, MOUNTING

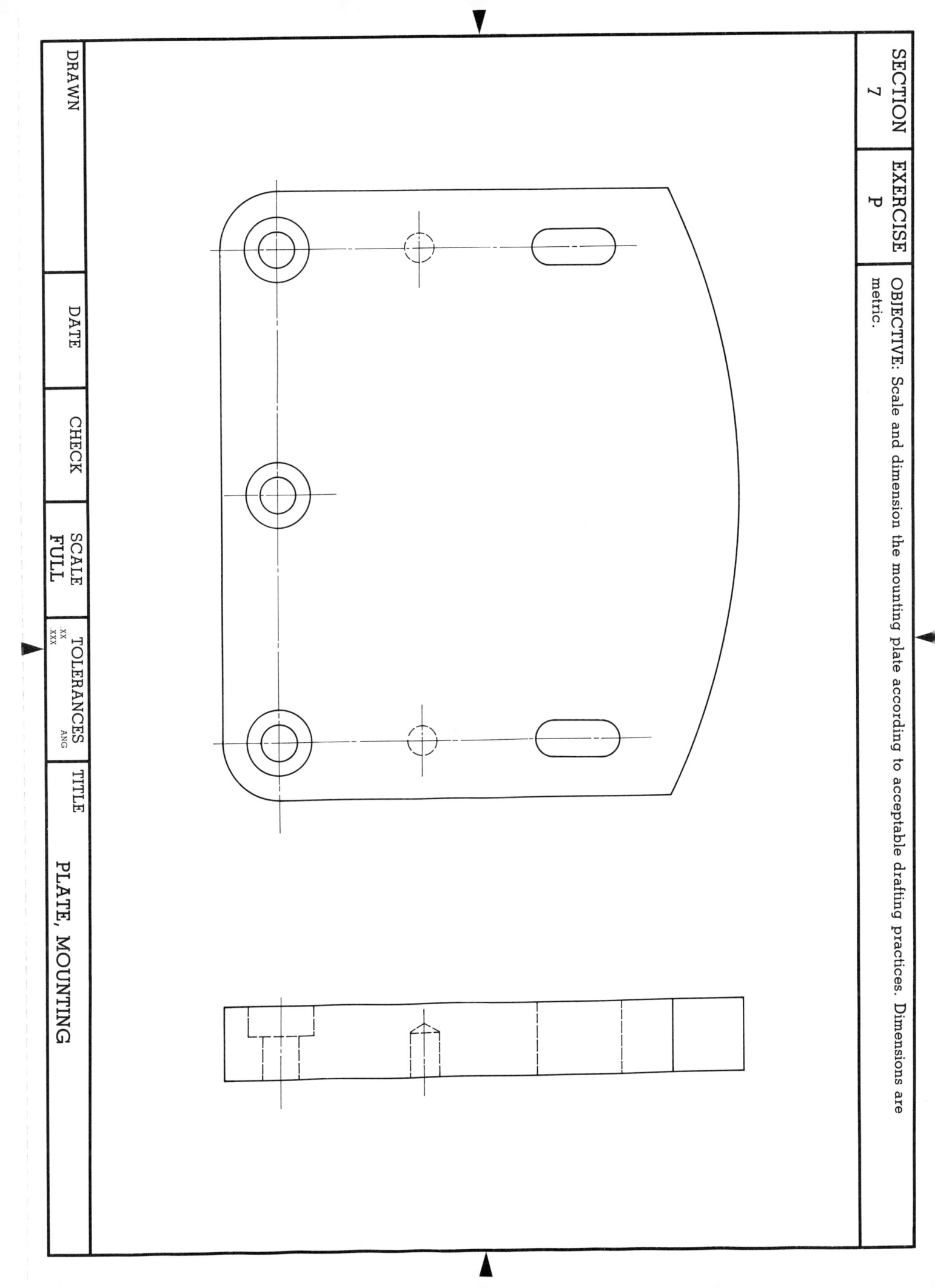

SECTION	EXERCISE
7	Q

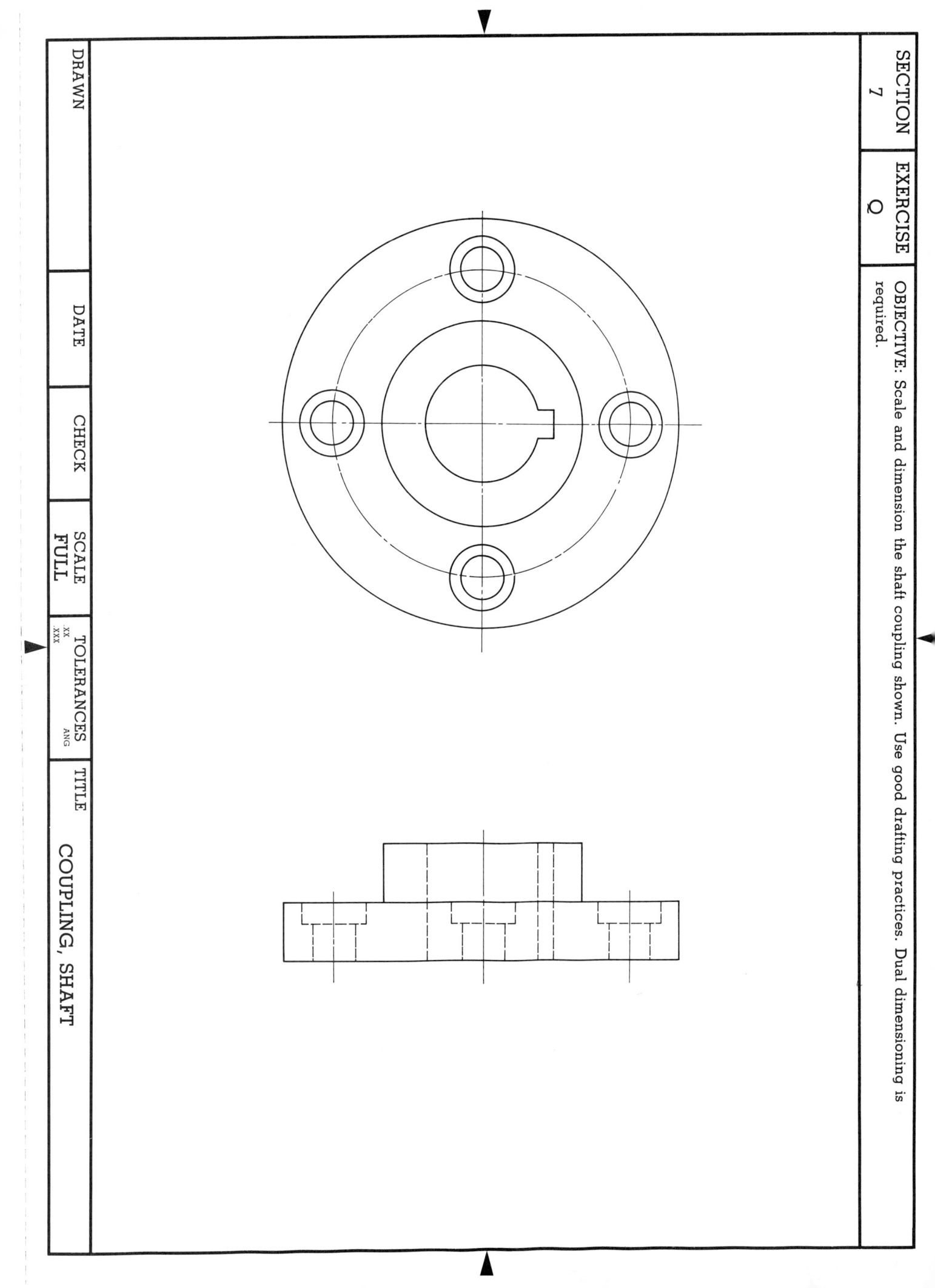

OBJECTIVE: Scale and dimension the shaft coupling shown. Use good drafting practices. Dual dimensioning is required.

DRAWN		
DATE	CHECK	
	SCALE FULL	
	TOLERANCES XX XXX ANG	TITLE
		COUPLING, SHAFT

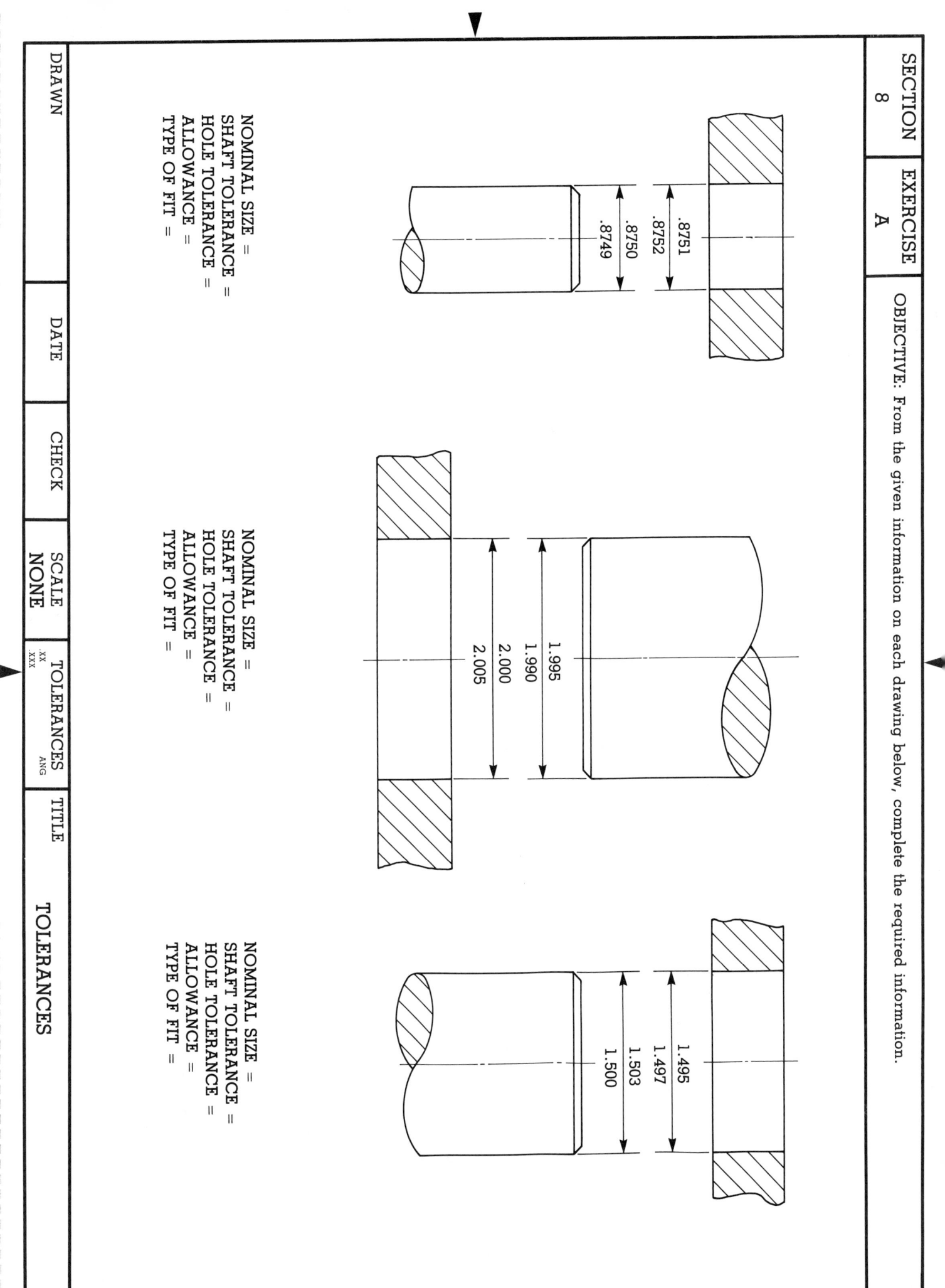

| SECTION 8 | EXERCISE A | OBJECTIVE: From the given information on each drawing below, complete the required information. |

.8751
.8752
.8750
.8749

NOMINAL SIZE =
SHAFT TOLERANCE =
HOLE TOLERANCE =
ALLOWANCE =
TYPE OF FIT =

1.995
1.990
2.000
2.005

NOMINAL SIZE =
SHAFT TOLERANCE =
HOLE TOLERANCE =
ALLOWANCE =
TYPE OF FIT =

1.495
1.497
1.503
1.500

NOMINAL SIZE =
SHAFT TOLERANCE =
HOLE TOLERANCE =
ALLOWANCE =
TYPE OF FIT =

| DRAWN | DATE | CHECK | SCALE NONE | TOLERANCES .XX .XXX ANG | TITLE |

TOLERANCES

SECTION 8	EXERCISE B	OBJECTIVE: Complete the table below for each case identified using correct limit dimensions.

	NOMINAL SIZE	BASIC SIZE	SHAFT TOLERANCE	HOLE TOLERANCE	ALLOWANCE	BASIC HOLE UNILATERAL	BASIC SHAFT UNILATERAL	BASIC HOLE BILATERAL	BASIC SHAFT BILATERAL	FIT CLEARANCE (C) INTERFER (I)
DECIMAL INCH VALUES	¾		.002	.002	.005					C
		1.000	.002	.002	.004	╳	╳			I
		1.500	.001	.001	.003	╳	╳		╳	C
		2.250	.0005	.0005	.002			╳	╳	I

SHOW MATH HERE

DRAWN	DATE	CHECK	SCALE	TOLERANCES .XX .XXX ANG	TITLE

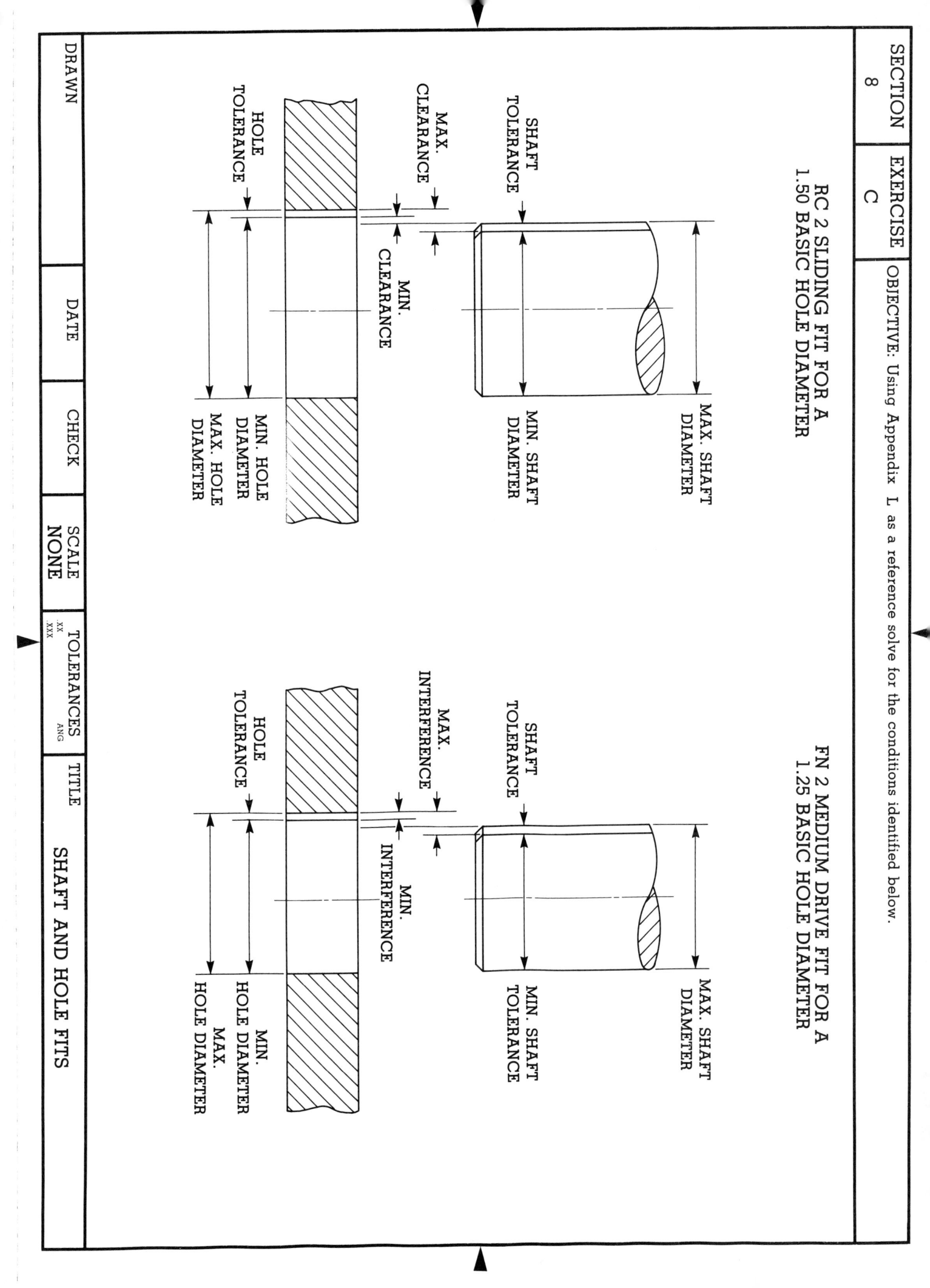

SECTION	EXERCISE	OBJECTIVE: Using Appendix L as a reference solve for the conditions identified below.
8	C	

RC 2 SLIDING FIT FOR A 1.50 BASIC HOLE DIAMETER

FN 2 MEDIUM DRIVE FIT FOR A 1.25 BASIC HOLE DIAMETER

DRAWN	DATE	CHECK	SCALE NONE	TOLERANCES .XX .XXX ANG	TITLE SHAFT AND HOLE FITS

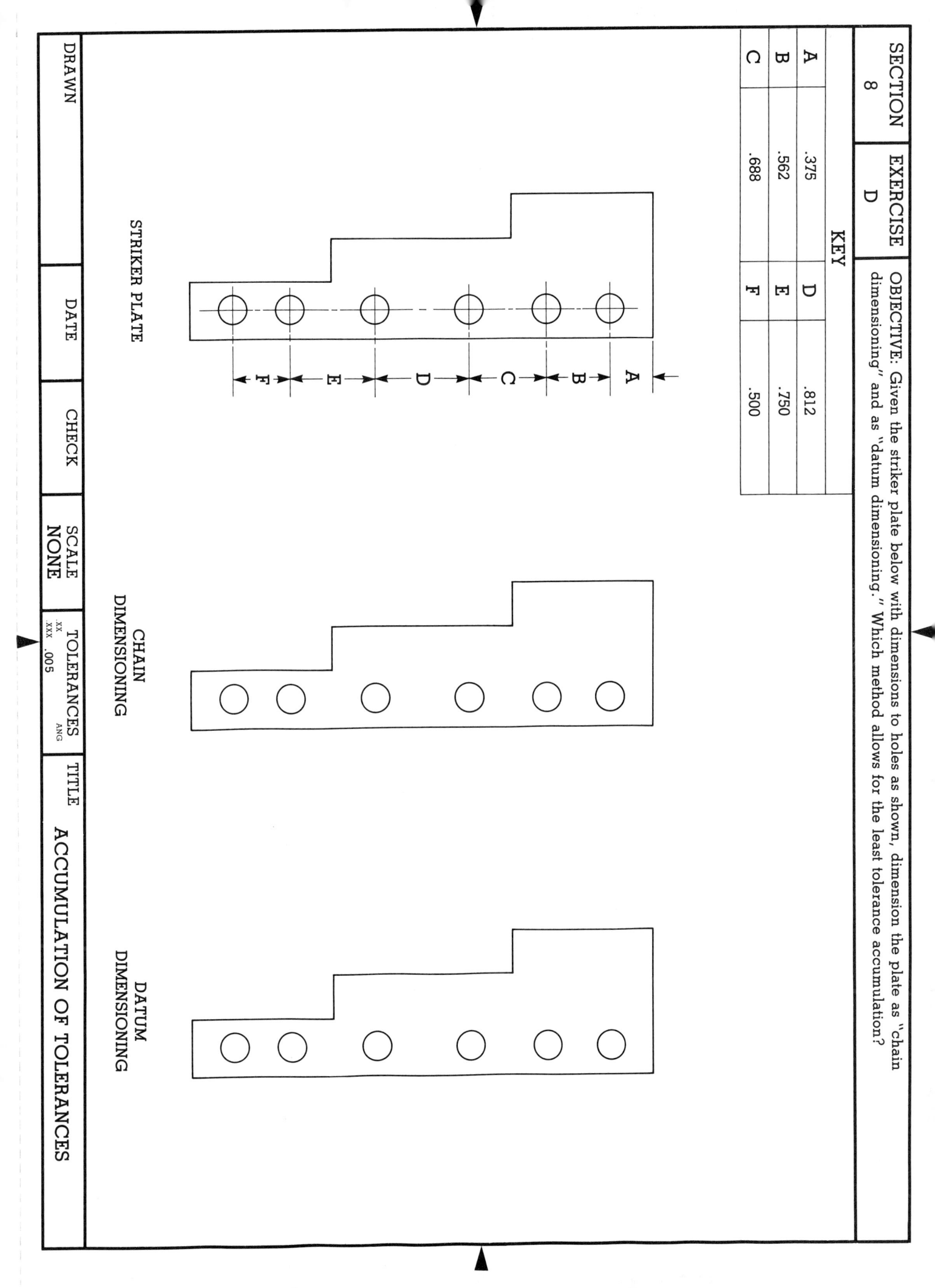

SECTION	EXERCISE	
8	D	OBJECTIVE: Given the striker plate below with dimensions to holes as shown, dimension the plate as "chain dimensioning" and as "datum dimensioning." Which method allows for the least tolerance accumulation?

KEY			
A	.375	D	.812
B	.562	E	.750
C	.688	F	.500

STRIKER PLATE

CHAIN DIMENSIONING

DATUM DIMENSIONING

| DRAWN | DATE | CHECK | SCALE NONE | TOLERANCES XX XXX .005 ANG | TITLE ACCUMULATION OF TOLERANCES |

SECTION 8 — EXERCISE E

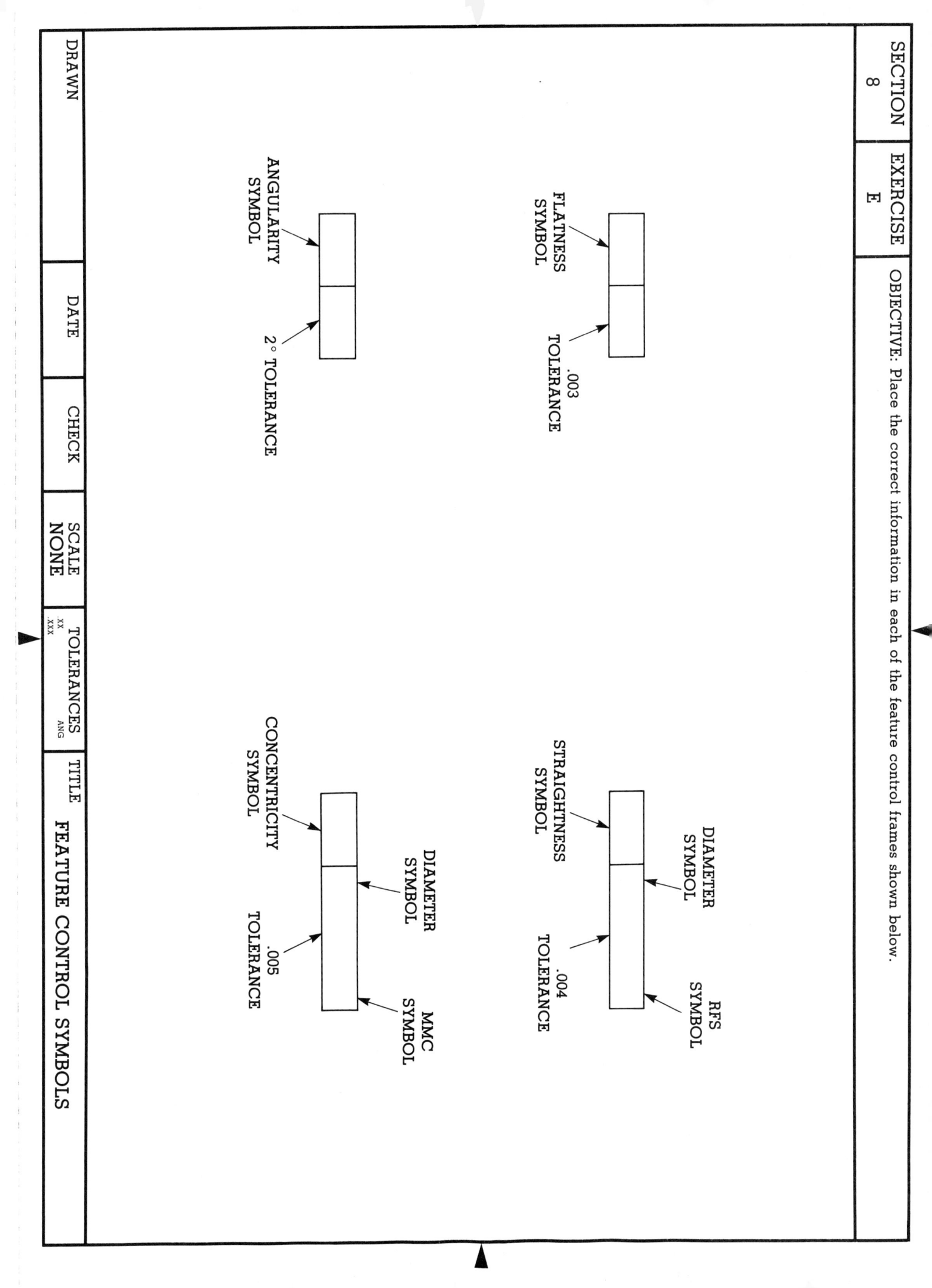

OBJECTIVE: Place the correct information in each of the feature control frames shown below.

FLATNESS SYMBOL → □□ ← .003 TOLERANCE

ANGULARITY SYMBOL → □□ ← 2° TOLERANCE

STRAIGHTNESS SYMBOL → □□□□
- DIAMETER SYMBOL
- .004 TOLERANCE
- RFS SYMBOL

CONCENTRICITY SYMBOL → □□□□
- DIAMETER SYMBOL
- .005 TOLERANCE
- MMC SYMBOL

| DRAWN | DATE | CHECK | SCALE NONE | TOLERANCES XX / XXX / ANG | TITLE FEATURE CONTROL SYMBOLS |

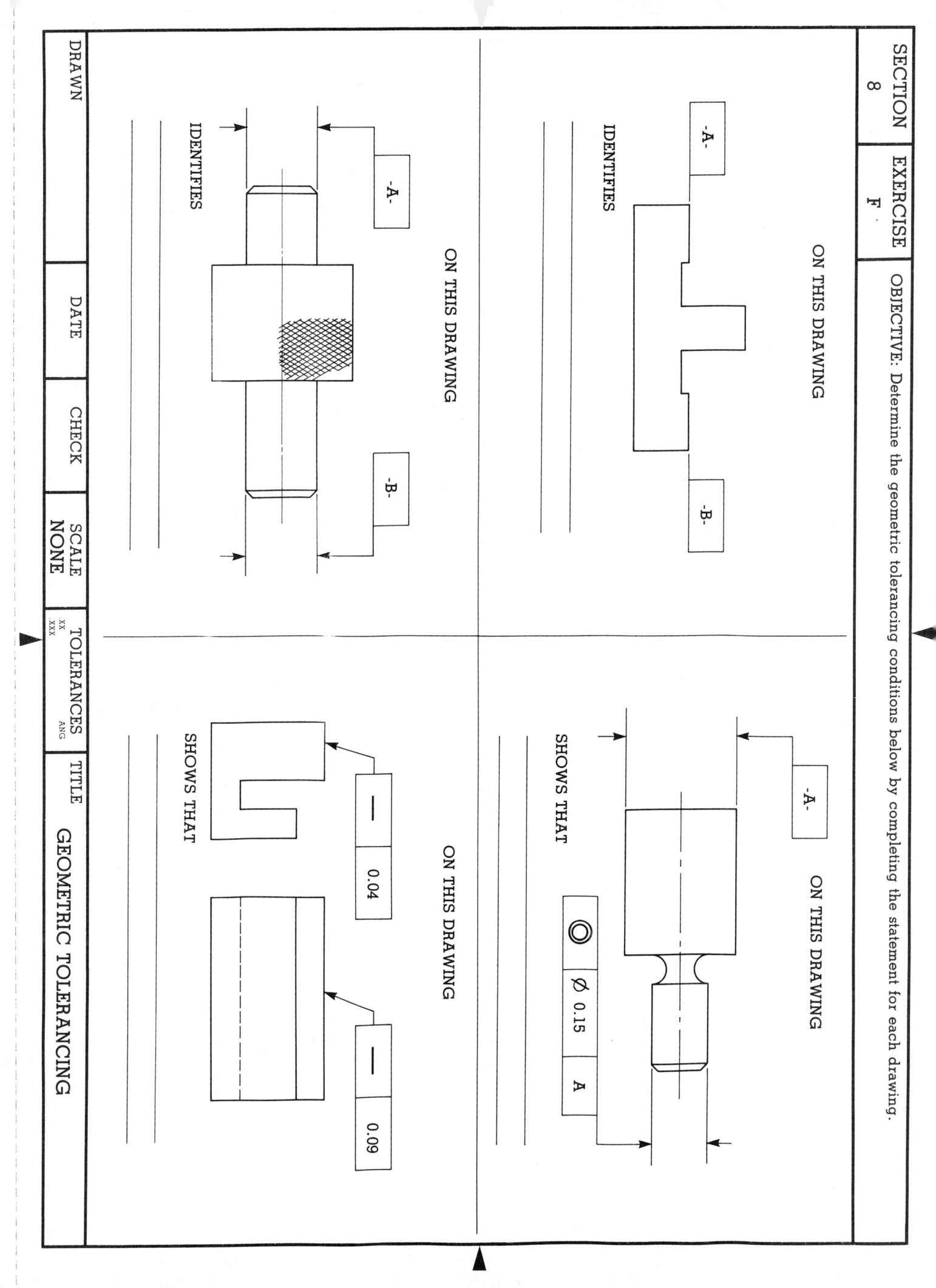

SECTION	EXERCISE	OBJECTIVE: Determine the geometric tolerancing conditions below by completing the statement for each drawing.
8	F	

-A-

IDENTIFIES _____ ON THIS DRAWING

-B-

-A-

IDENTIFIES _____ ON THIS DRAWING

-B-

-A-

SHOWS THAT _____ ON THIS DRAWING

◎ | ⌀ 0.15 | A

SHOWS THAT _____ ON THIS DRAWING

— | 0.04

— | 0.09

| DRAWN | DATE | CHECK | SCALE NONE | TOLERANCES ANG XX XXX | TITLE GEOMETRIC TOLERANCING |

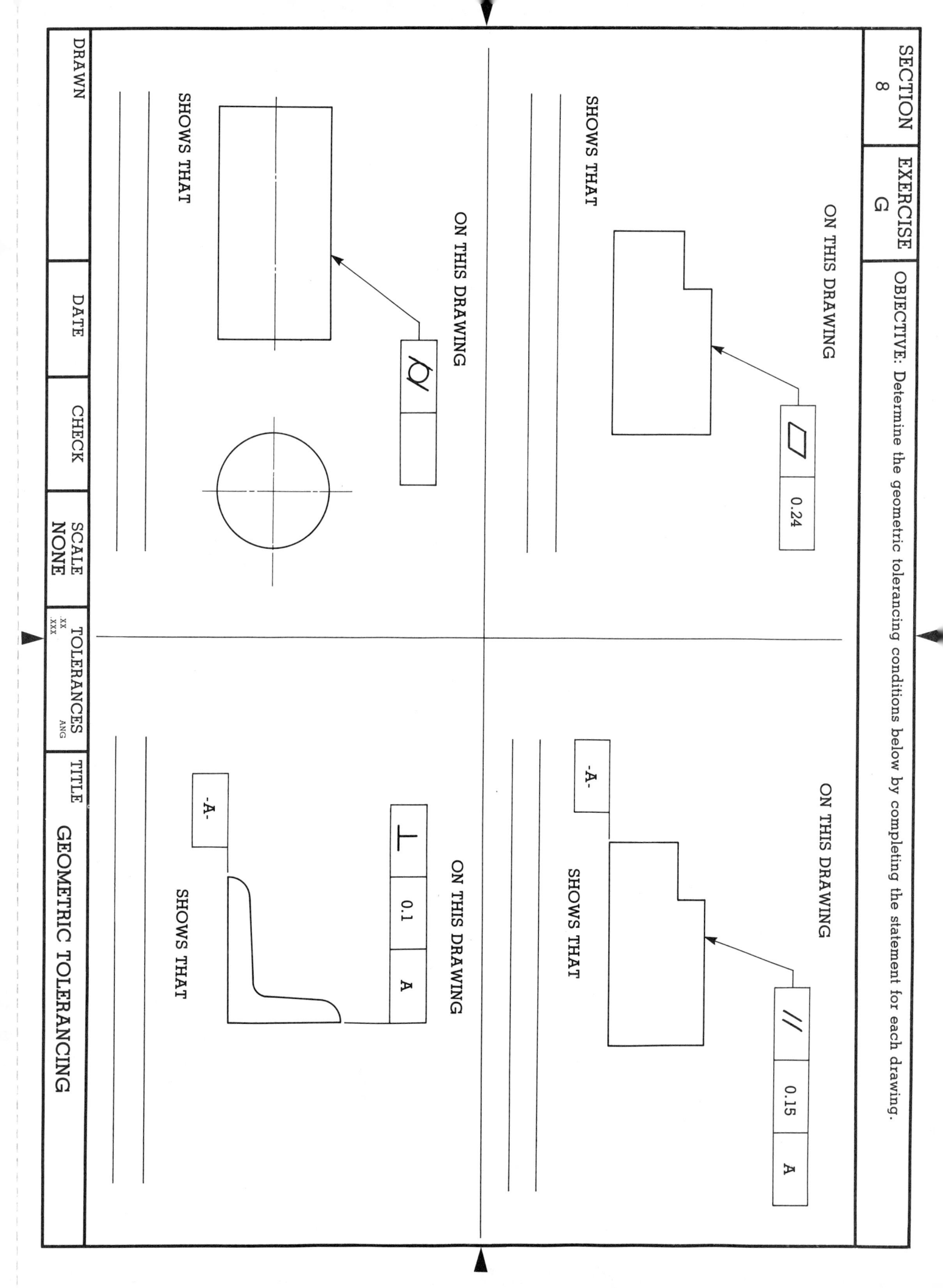

| SECTION 8 | EXERCISE G | OBJECTIVE: Determine the geometric tolerancing conditions below by completing the statement for each drawing. |

ON THIS DRAWING

SHOWS THAT
_____
_____
_____

ON THIS DRAWING

SHOWS THAT
_____
_____
_____

ON THIS DRAWING

SHOWS THAT
_____
_____
_____

ON THIS DRAWING

SHOWS THAT
_____
_____
_____

| DRAWN | DATE | CHECK | SCALE NONE | TOLERANCES XX XXX ANG | TITLE GEOMETRIC TOLERANCING |

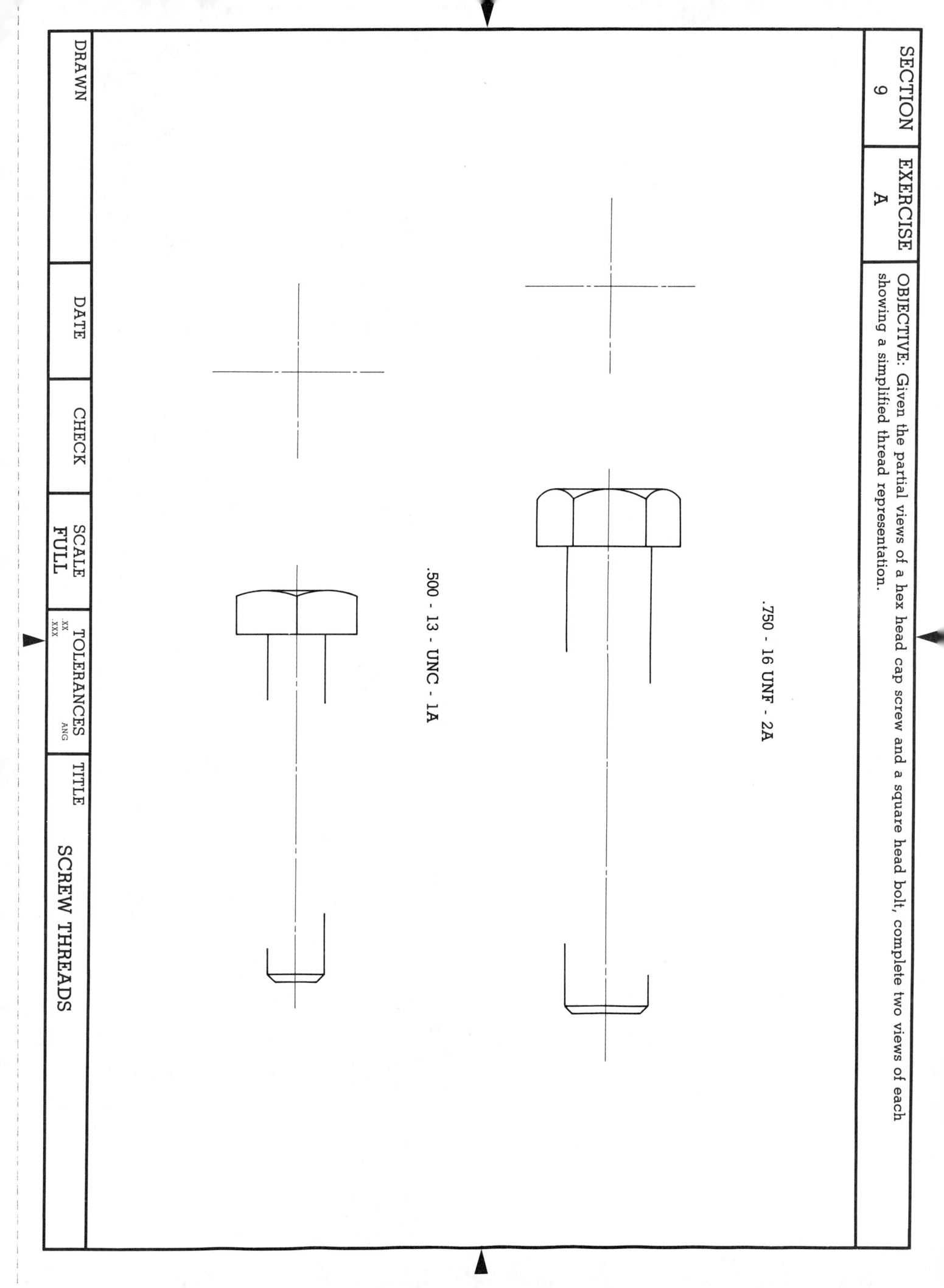

SECTION	EXERCISE		
9	A	**OBJECTIVE:** Given the partial views of a hex head cap screw and a square head bolt, complete two views of each showing a simplified thread representation.	

.750 - 16 UNF - 2A

.500 - 13 - UNC - 1A

DRAWN	DATE	CHECK	SCALE FULL	TOLERANCES XX / XXX / ANG	TITLE
					SCREW THREADS

SECTION	EXERCISE	
9	B	OBJECTIVE: Draw each of the hardware items in the space provided below. Show simplified thread representation.

#10 - 24 x .75 LG. ROUND HD MACHINE SCREW

#8 - 32 x 1.00 LG. FLAT HD MACHINE SCREW

#10 - 32 HEX NUT
(2 VIEWS)

.250 - 20 x 1.25 LG. HEX HD CAP SCREW

¼ PLAIN WASHER
(2 VIEWS)

DRAWN	DATE	CHECK	SCALE 2:1	TOLERANCES XX XXX ANG	TITLE
					HARDWARE

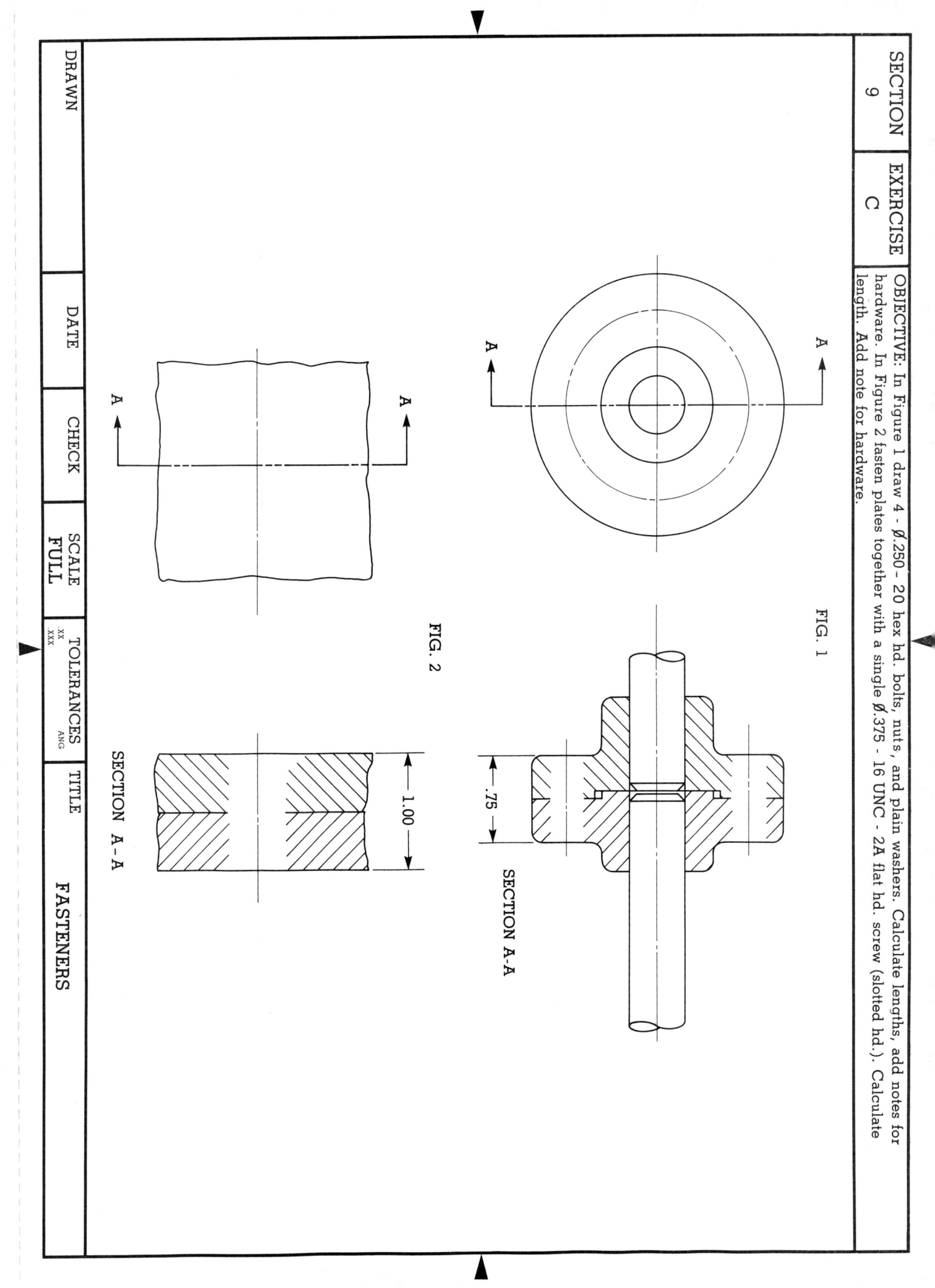

SECTION	EXERCISE					TITLE
9	C	DRAWN	DATE	CHECK	SCALE FULL	FASTENERS

TOLERANCES: XX, XXX, ANG

OBJECTIVE: In Figure 1 draw 4 - ⌀.250 - 20 hex hd. bolts, nuts, and plain washers. Calculate lengths, add notes for hardware. In Figure 2 fasten plates together with a single ⌀.375 - 16 UNC - 2A flat hd. screw (slotted hd.). Calculate length. Add note for hardware.

FIG. 1

SECTION A-A

.75

FIG. 2

SECTION A-A

1.00

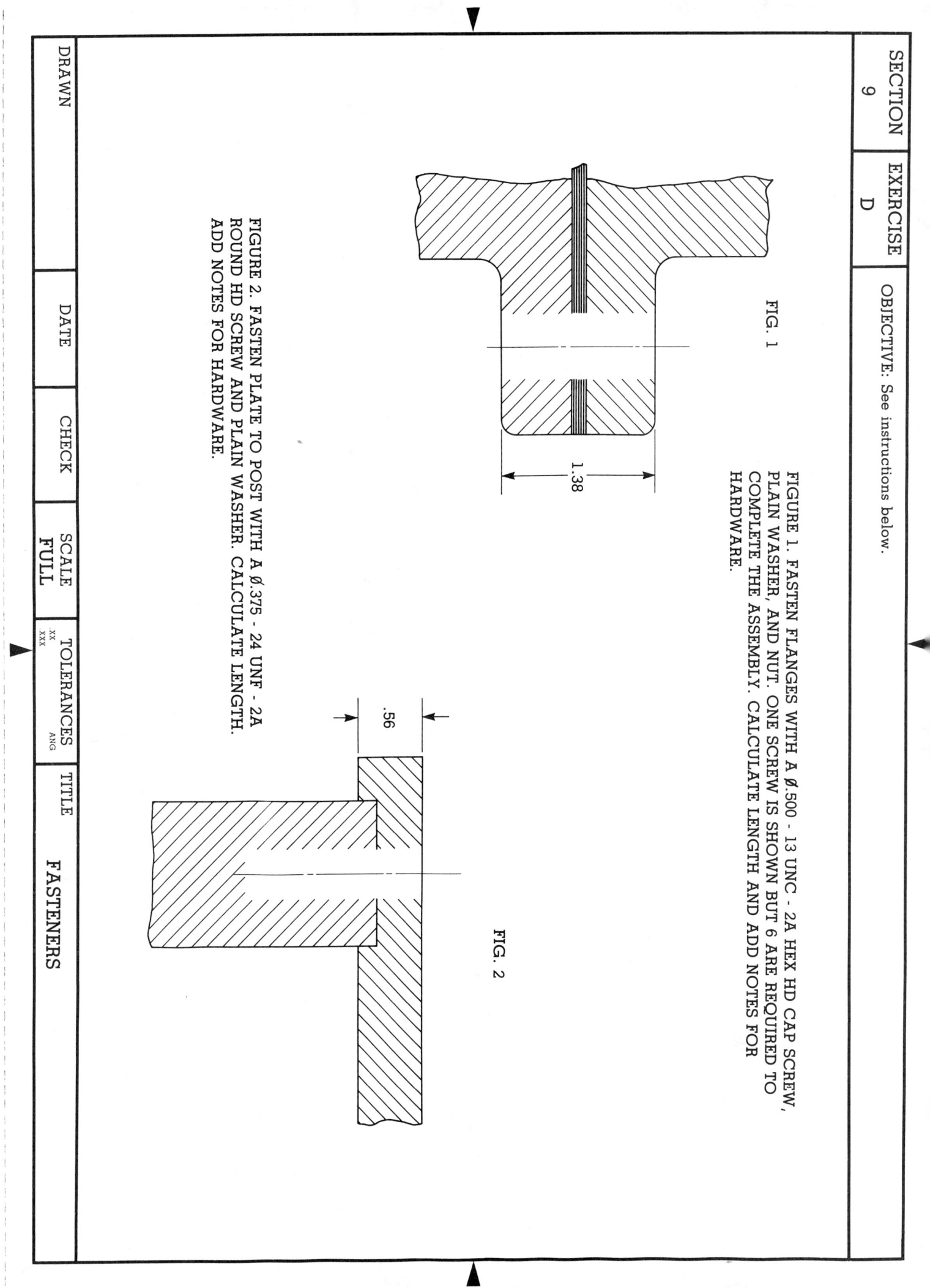

SECTION	EXERCISE	OBJECTIVE: See instructions below.
9	D	

FIG. 1

1.38

FIGURE 1. FASTEN FLANGES WITH A ⌀.500 - 13 UNC - 2A HEX HD CAP SCREW, PLAIN WASHER, AND NUT. ONE SCREW IS SHOWN BUT 6 ARE REQUIRED TO COMPLETE THE ASSEMBLY. CALCULATE LENGTH AND ADD NOTES FOR HARDWARE.

FIG. 2

.56

FIGURE 2. FASTEN PLATE TO POST WITH A ⌀.375 - 24 UNF - 2A ROUND HD SCREW AND PLAIN WASHER. CALCULATE LENGTH. ADD NOTES FOR HARDWARE.

DRAWN	DATE	CHECK	SCALE FULL	TOLERANCES XX XXX ANG	TITLE FASTENERS

SECTION	EXERCISE	OBJECTIVE: See instructions below.
9	E	

FIGURE 1. DRAW 3 VIEWS OF A THREADED STUD. ADD DIMENSIONS AND MANUFACTURING NOTES. THE LARGE DIAMETER IS ⌀1.00 - 8 UNC - 2A AND IS 2.50 INCHES LONG. THE SMALLER END IS ⌀.625 - 11 UNC - 2A. OVERALL LENGTH IS 4.00 INCHES.

FIG. 1

⌀.75

.75

FIG. 2

FIGURE 2. A PIN IS ATTACHED TO A STEEL PLATE WHOSE THREADED PORTION HAS A ⌀.375 - 16 UNC - 2A THREAD AND MOUNTS THROUGH A CLEARANCE HOLE WITH A HEX HUT AND PLAIN WASHER. CALCULATE LENGTH AND ADD MANUFACTURING NOTES.

DRAWN	DATE	CHECK	SCALE FULL	TOLERANCES XX XXX ANG	TITLE
					FASTENERS

SECTION	EXERCISE	OBJECTIVE: See instructions below.
9	F	

GIVEN: 4 - #4-40 SCREWS THAT MOUNT THROUGH CLEARANCE HOLES, DETERMINE MINIMUM SIZE OF CLEARANCE HOLES IF THE BILATERAL TOLERANCE BETWEEN HOLES IS ±.005. SHOW FULL LENGTH OF THREADS ON SCREWS.

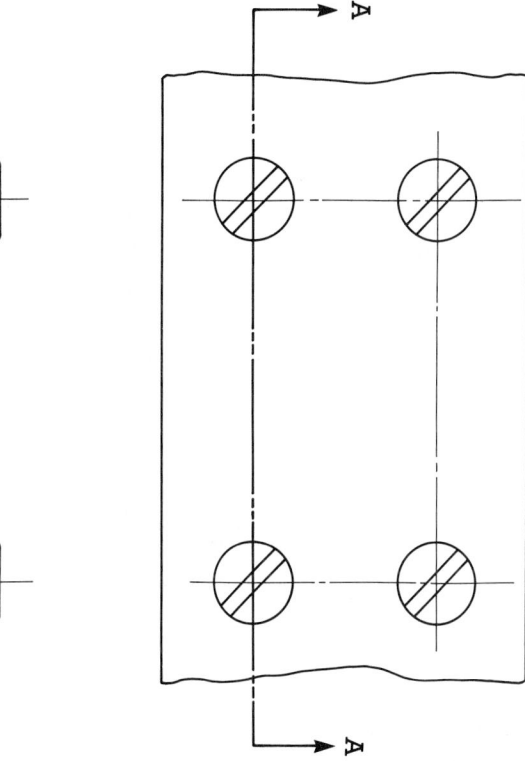

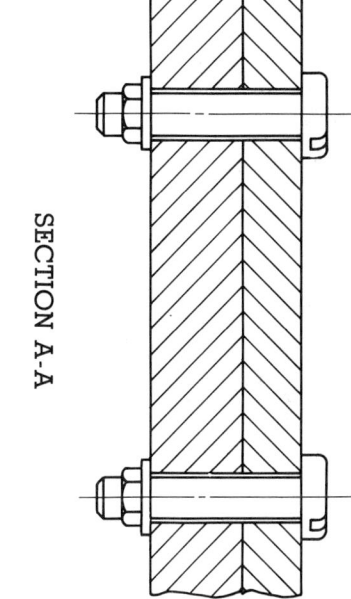

SECTION A-A

CLEARANCE HOLE DIA. =

GIVEN: A PATTERN OF 3 - #10-24 SCREWS FASTENING TWO PLATES (ONE WITH CLEARANCE HOLES, ONE WITH TAPPED HOLES), COMPLETE THE DRAWING THEN DETERMINE MINIMUM SIZE OF CLEARANCE HOLES IF THE BILATERAL TOLERANCE IS ±.007.

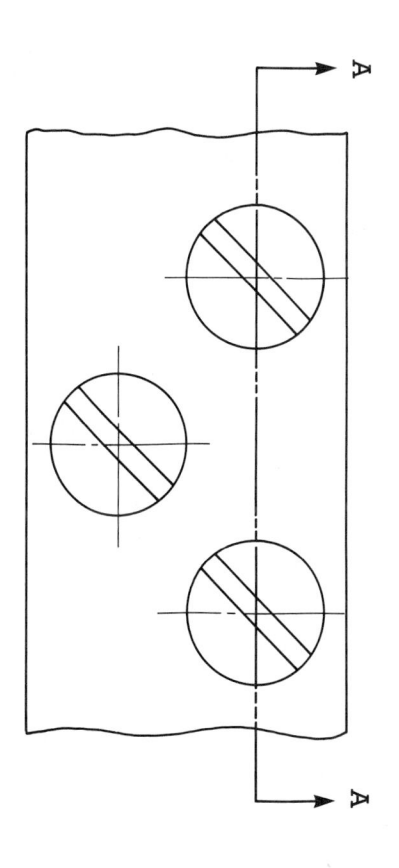

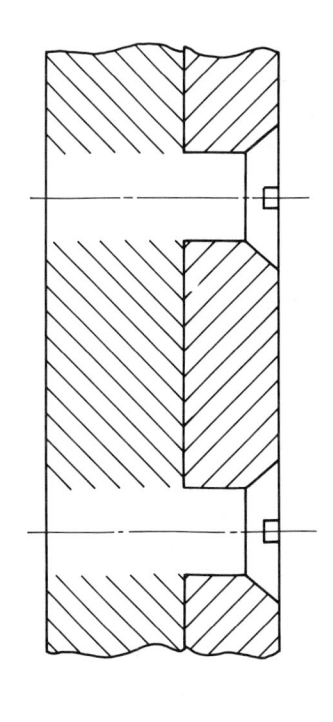

SECTION A-A

CLEARANCE HOLE DIA. =

DRAWN	DATE	CHECK	SCALE 2:1	TOLERANCES .XX .XXX ANG	TITLE FASTENERS

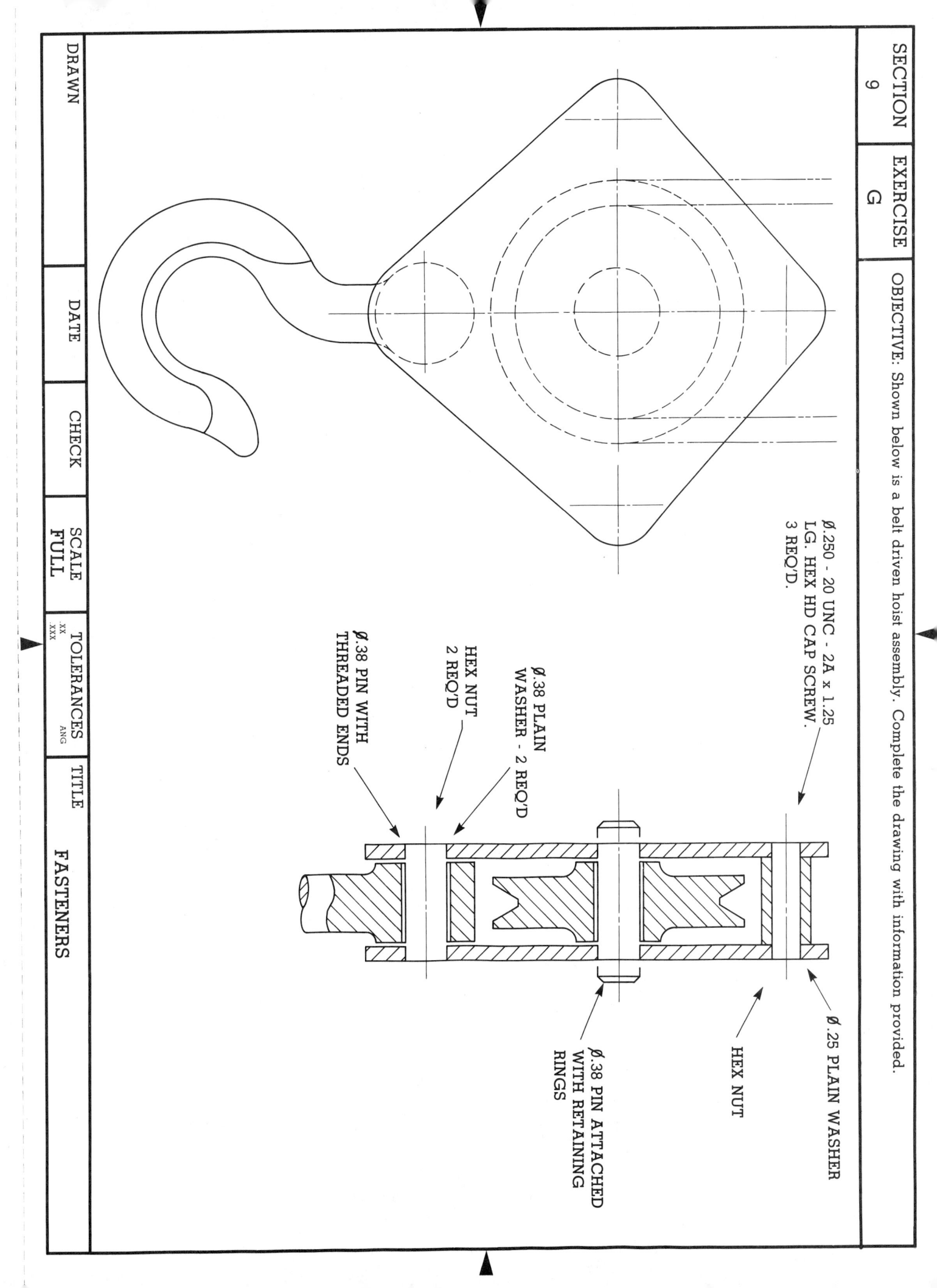

SECTION	EXERCISE	OBJECTIVE: Shown below is a belt driven hoist assembly. Complete the drawing with information provided.
9	G	

Ø.250 - 20 UNC - 2A x 1.25 LG. HEX HD CAP SCREW. 3 REQ'D.

Ø.38 PIN WITH THREADED ENDS

HEX NUT 2 REQ'D

Ø.38 PLAIN WASHER - 2 REQ'D

Ø.25 PLAIN WASHER

HEX NUT

Ø.38 PIN ATTACHED WITH RETAINING RINGS

DRAWN	DATE	CHECK	SCALE FULL	TOLERANCES .XX .XXX ANG	TITLE
					FASTENERS

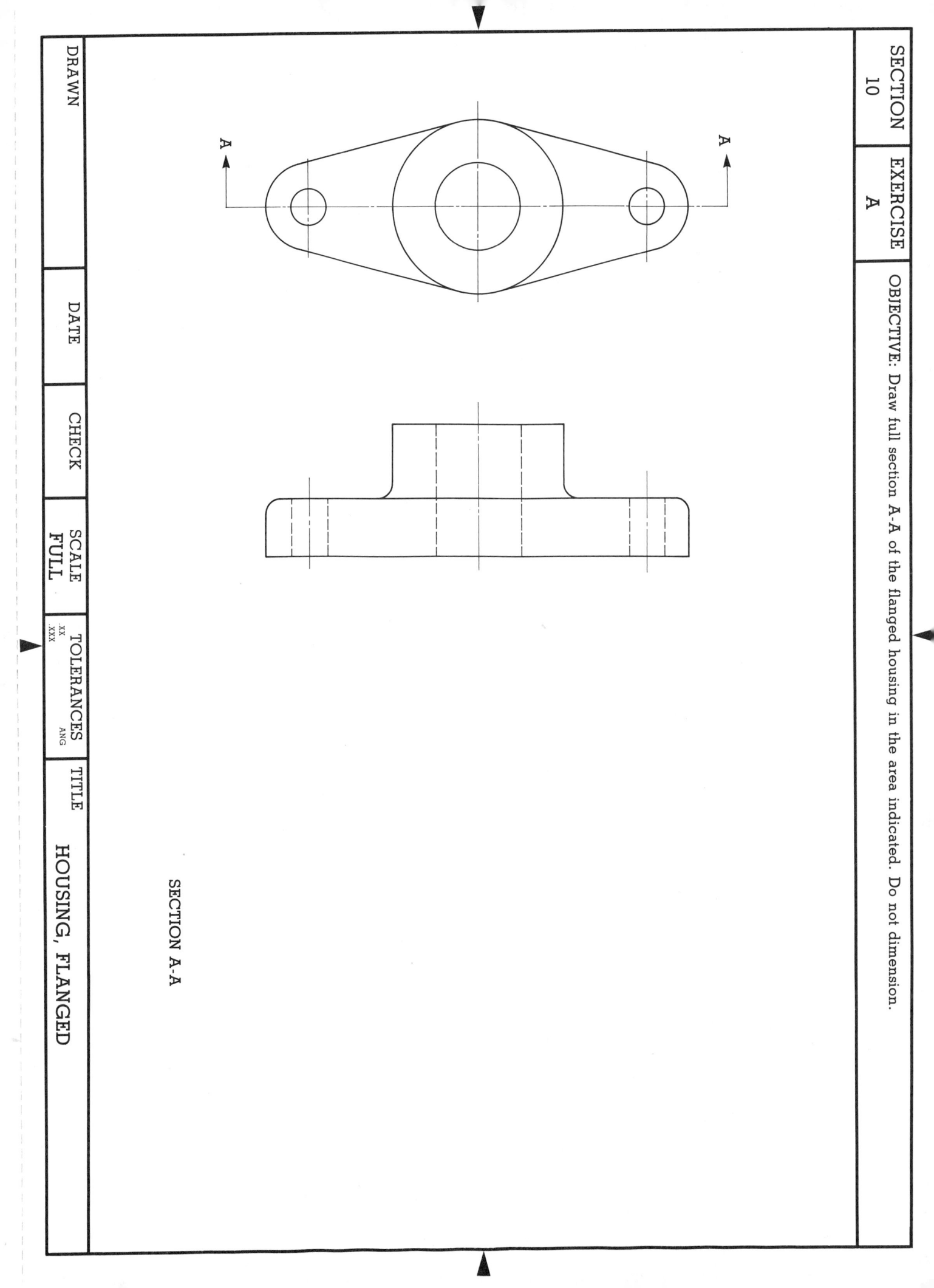

SECTION	EXERCISE	OBJECTIVE: Draw full section A-A of the flanged housing in the area indicated. Do not dimension.
10	A	

SECTION A-A

TITLE: HOUSING, FLANGED

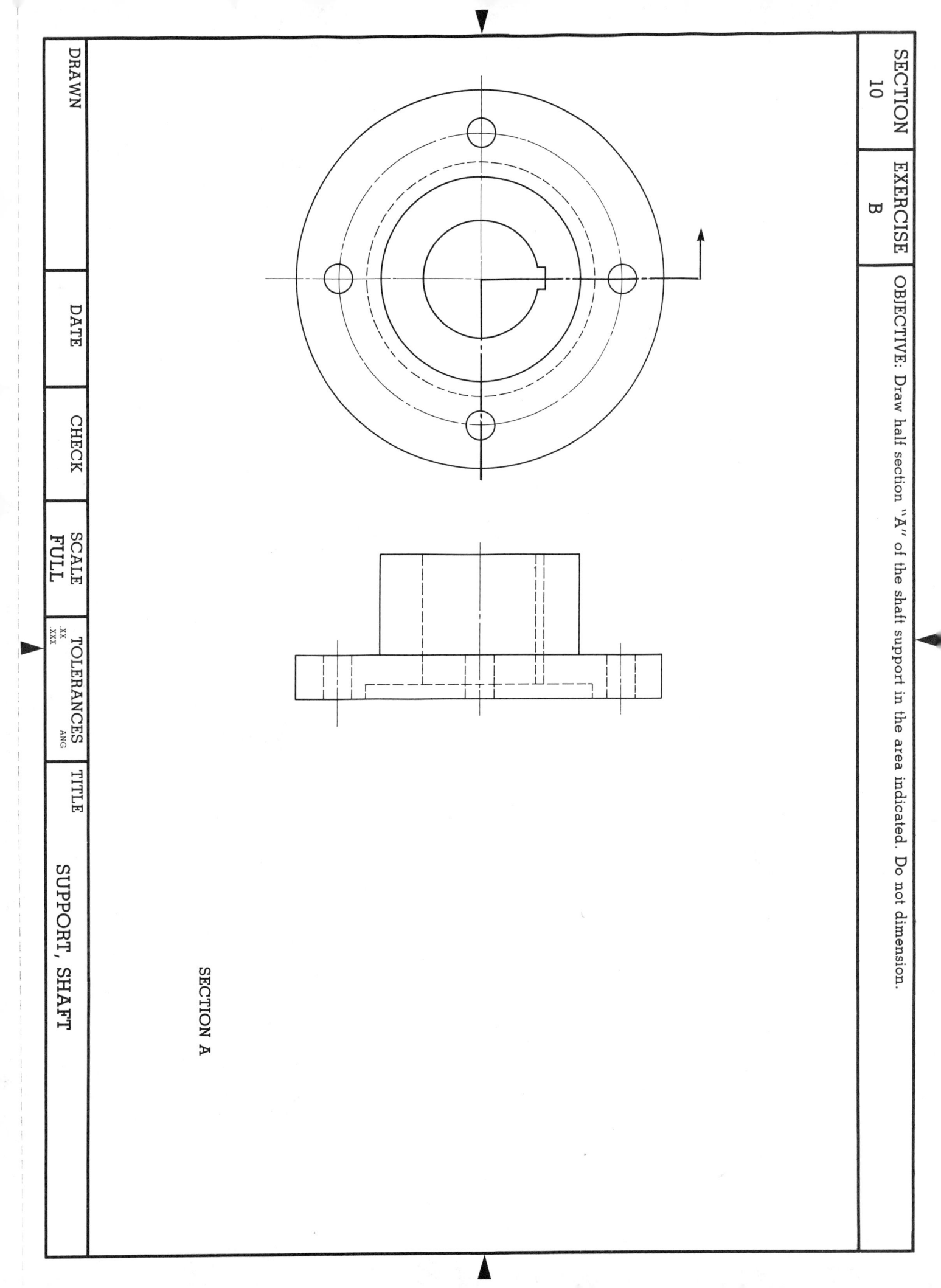

SECTION	EXERCISE	OBJECTIVE: Draw half section "A" of the shaft support in the area indicated. Do not dimension.
10	B	

SECTION A

DRAWN	DATE	CHECK	SCALE FULL	TOLERANCES xx / xxx / ANG	TITLE SUPPORT, SHAFT

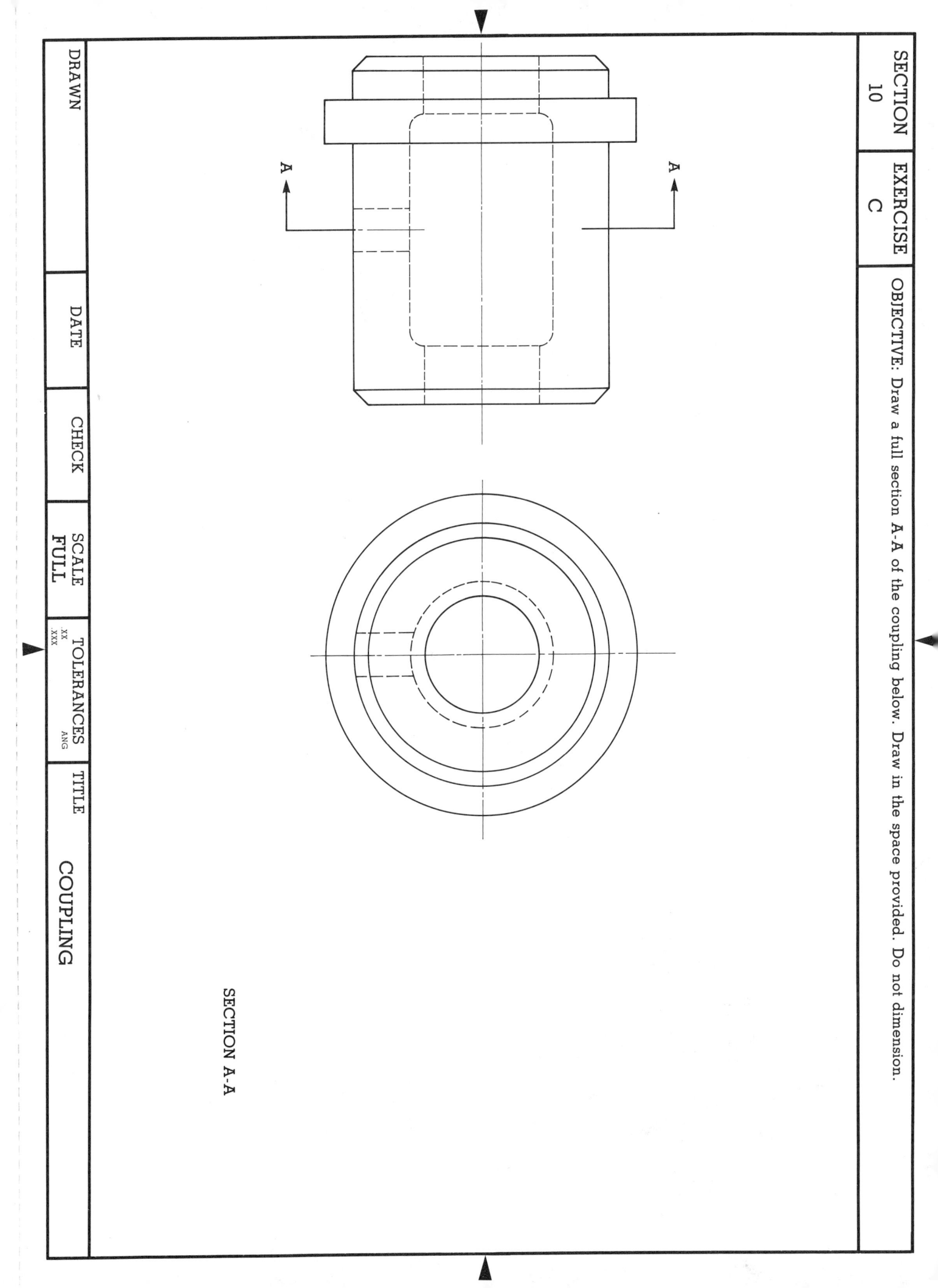

SECTION	EXERCISE	OBJECTIVE: Draw a full section A-A of the coupling below. Draw in the space provided. Do not dimension.
10	C	

SECTION A-A

DRAWN	DATE	CHECK	SCALE FULL	TOLERANCES XX XXX ANG	TITLE COUPLING

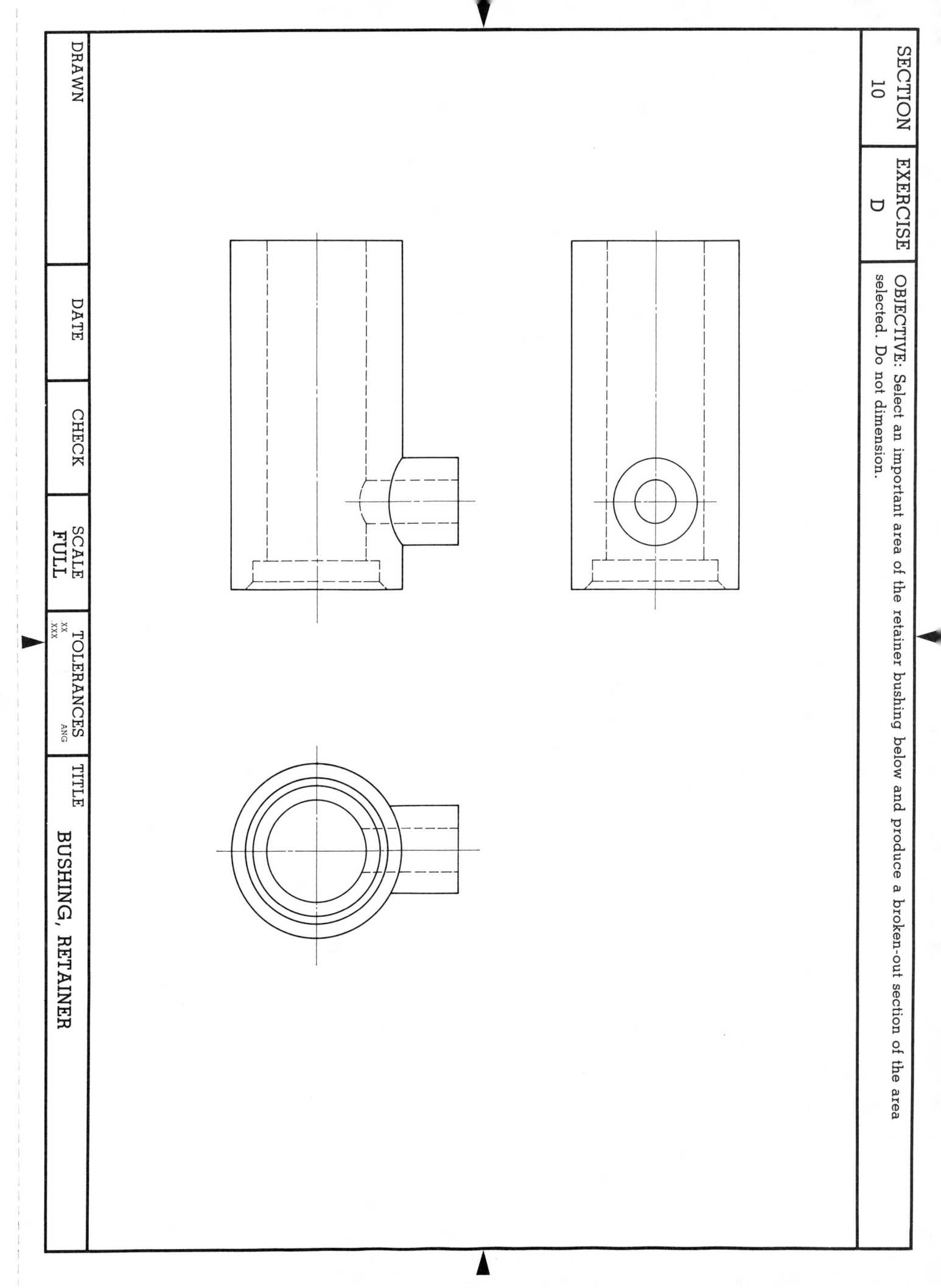

SECTION	EXERCISE	
10	D	OBJECTIVE: Select an important area of the retainer bushing below and produce a broken-out section of the area selected. Do not dimension.

DRAWN	DATE			
	CHECK	SCALE **FULL**	TOLERANCES XX ANG XXX	TITLE **BUSHING, RETAINER**

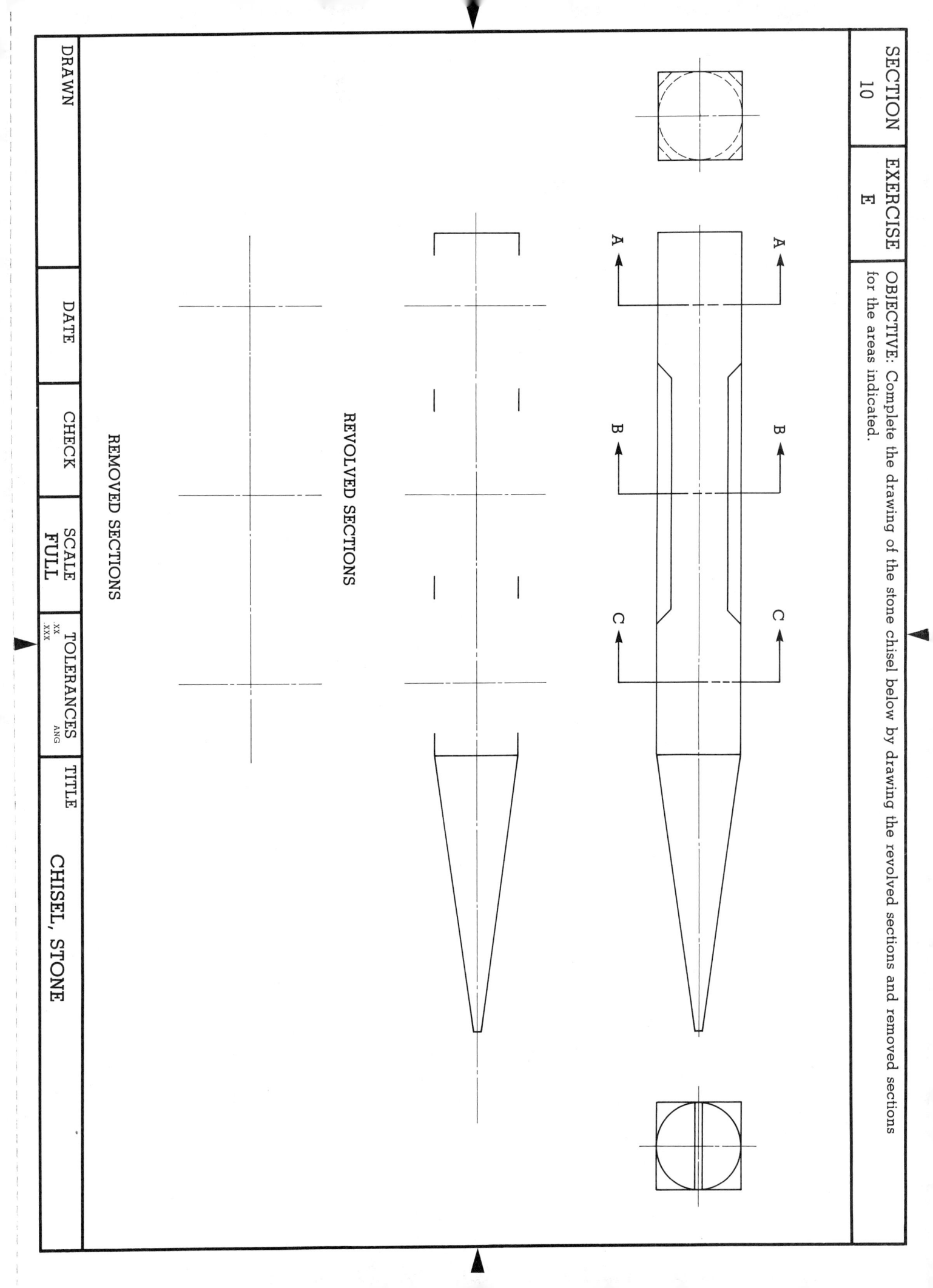

SECTION	EXERCISE	OBJECTIVE: Complete the drawing of the stone chisel below by drawing the revolved sections and removed sections for the areas indicated.
10	E	

REVOLVED SECTIONS

REMOVED SECTIONS

DRAWN	DATE	CHECK	SCALE FULL	TOLERANCES XX XXX ANG	TITLE
					CHISEL, STONE

SECTION	EXERCISE	OBJECTIVE: Scale the bearing housing below. Produce offset Section A-A in the indicated space.
10	F	

DRAWN	DATE	CHECK	SCALE FULL	TOLERANCES XX / XXX / ANG	TITLE
					HOUSING, BEARING

SECTION A-A

SECTION	EXERCISE
10	G

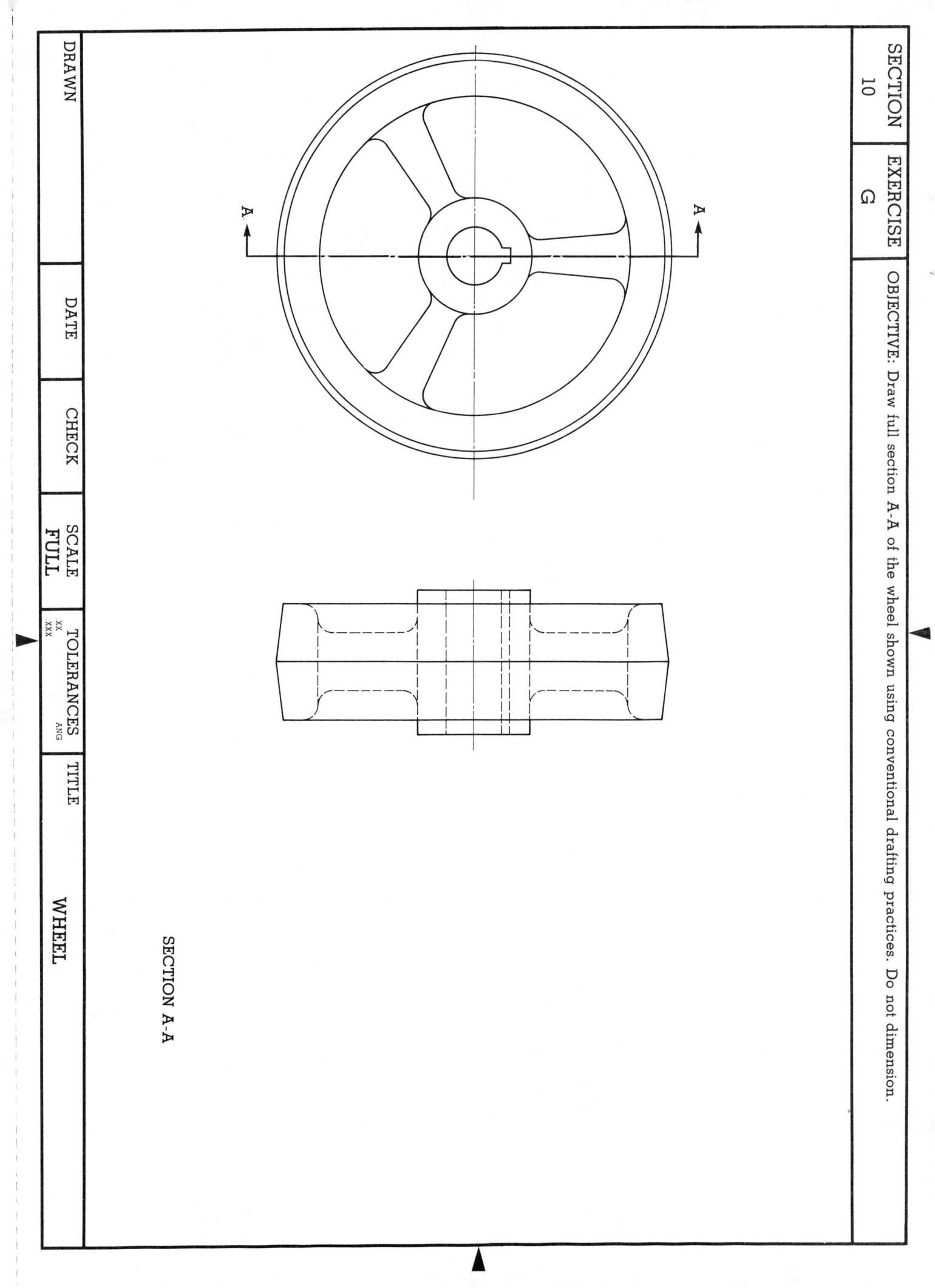

OBJECTIVE: Draw full section A-A of the wheel shown using conventional drafting practices. Do not dimension.

SECTION A-A

DRAWN	DATE	CHECK	SCALE FULL	TOLERANCES XX .XXX ANG	TITLE WHEEL

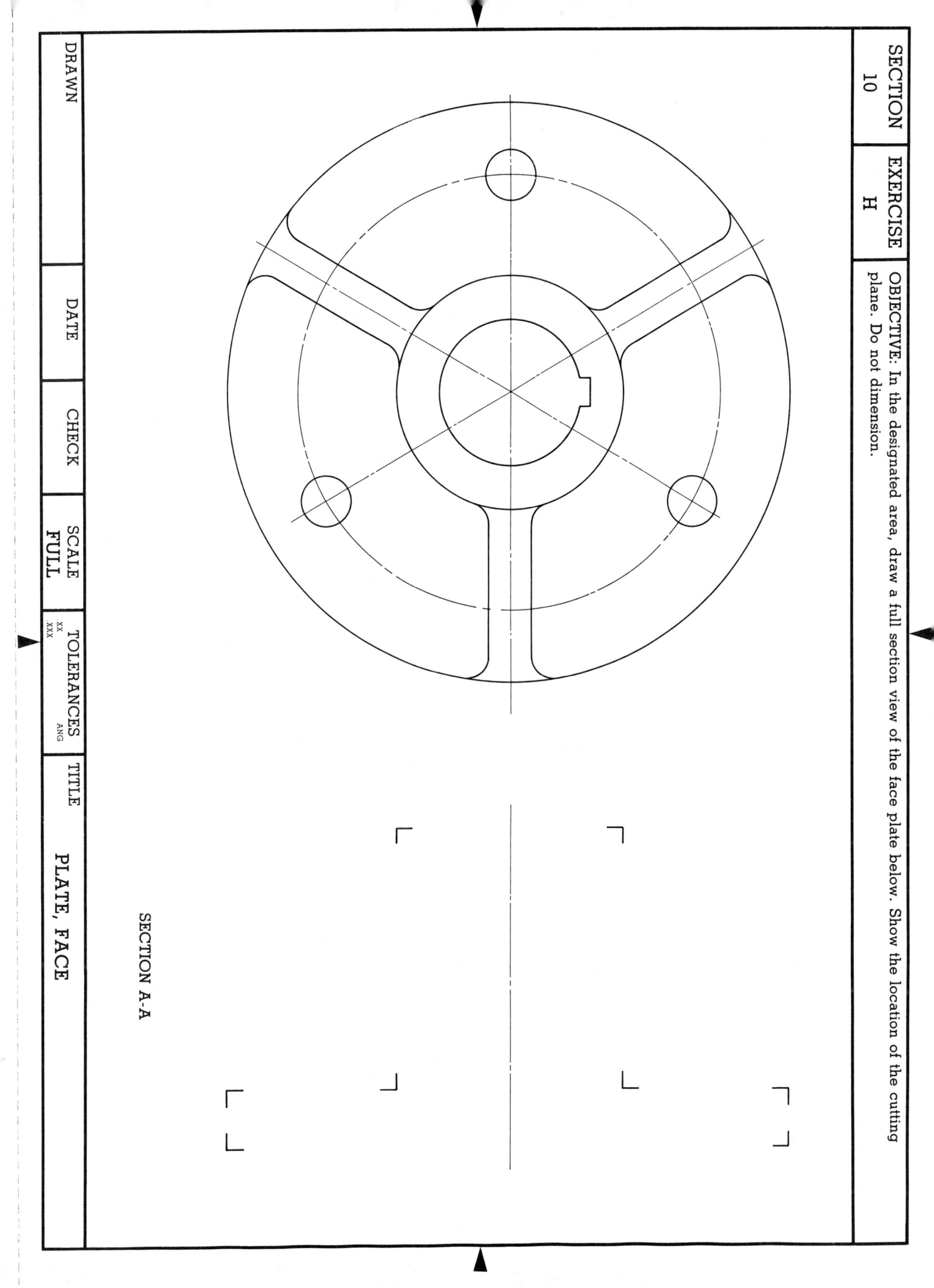

SECTION	EXERCISE	OBJECTIVE: In the designated area, draw a full section view of the face plate below. Show the location of the cutting plane. Do not dimension.
10	H	

SECTION A-A

DRAWN	DATE	SCALE FULL	TOLERANCES XX XXX ANG	TITLE
CHECK				PLATE, FACE

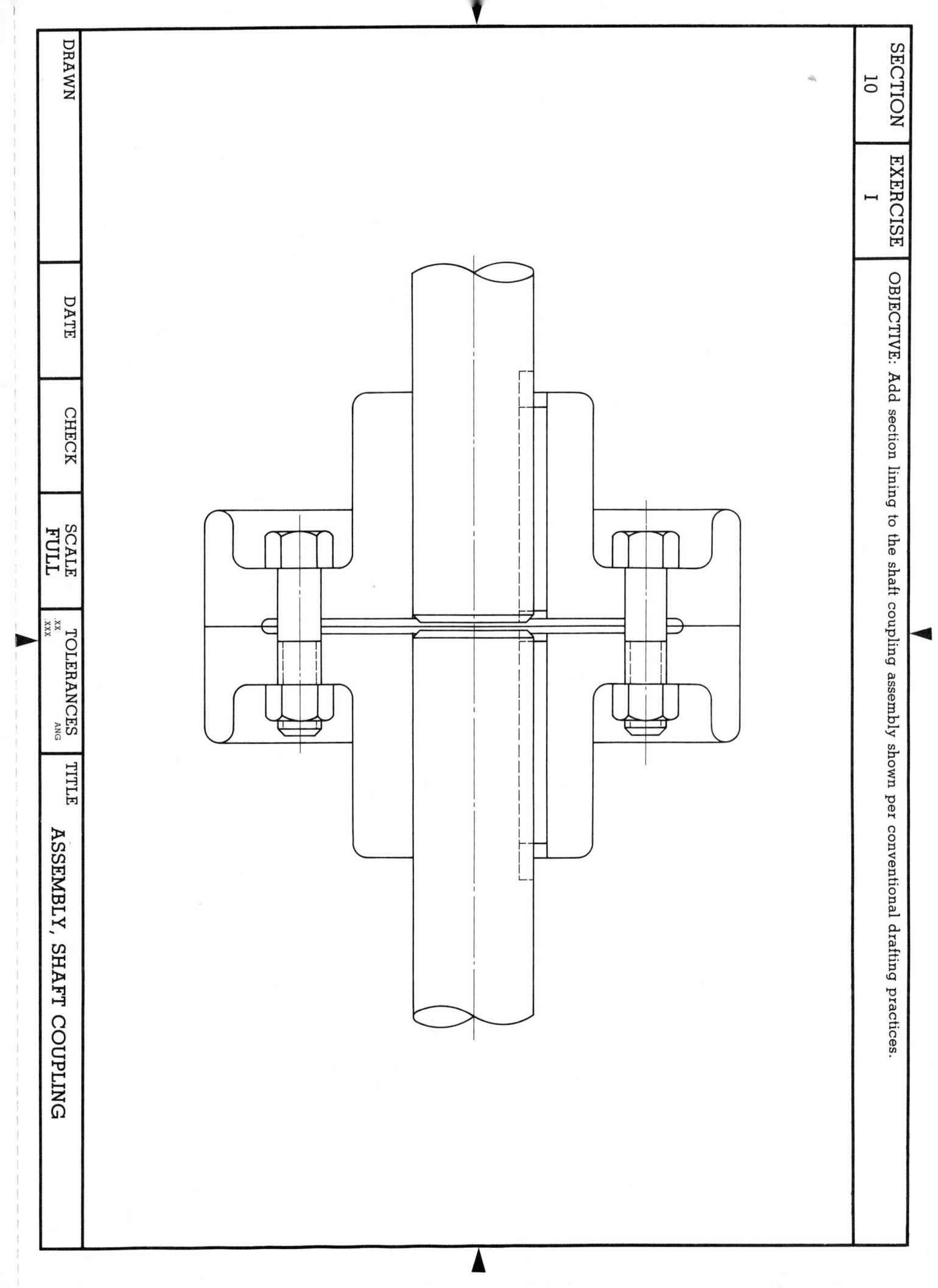

SECTION	EXERCISE	OBJECTIVE: Add section lining to the shaft coupling assembly shown per conventional drafting practices.
10	1	

DRAWN	DATE	CHECK	SCALE FULL	TOLERANCES .XX .XXX ANG	TITLE ASSEMBLY, SHAFT COUPLING

SECTION	EXERCISE	OBJECTIVE: Draw the front auxiliary view of surface "A" of the object shown below. Do not erase construction lines.
11	A	

DRAWN	DATE	CHECK	SCALE FULL	TOLERANCES XX XXX ANG	TITLE FRONT AUXILIARY VIEW

SECTION	EXERCISE	OBJECTIVE: Produce front auxiliary views for surfaces "A" and "B" for the object shown. Do not erase construction lines.
11	B	

DRAWN	DATE	CHECK	SCALE FULL	TOLERANCES XX / XXX / ANG	TITLE **FRONT AUXILIARY VIEWS**

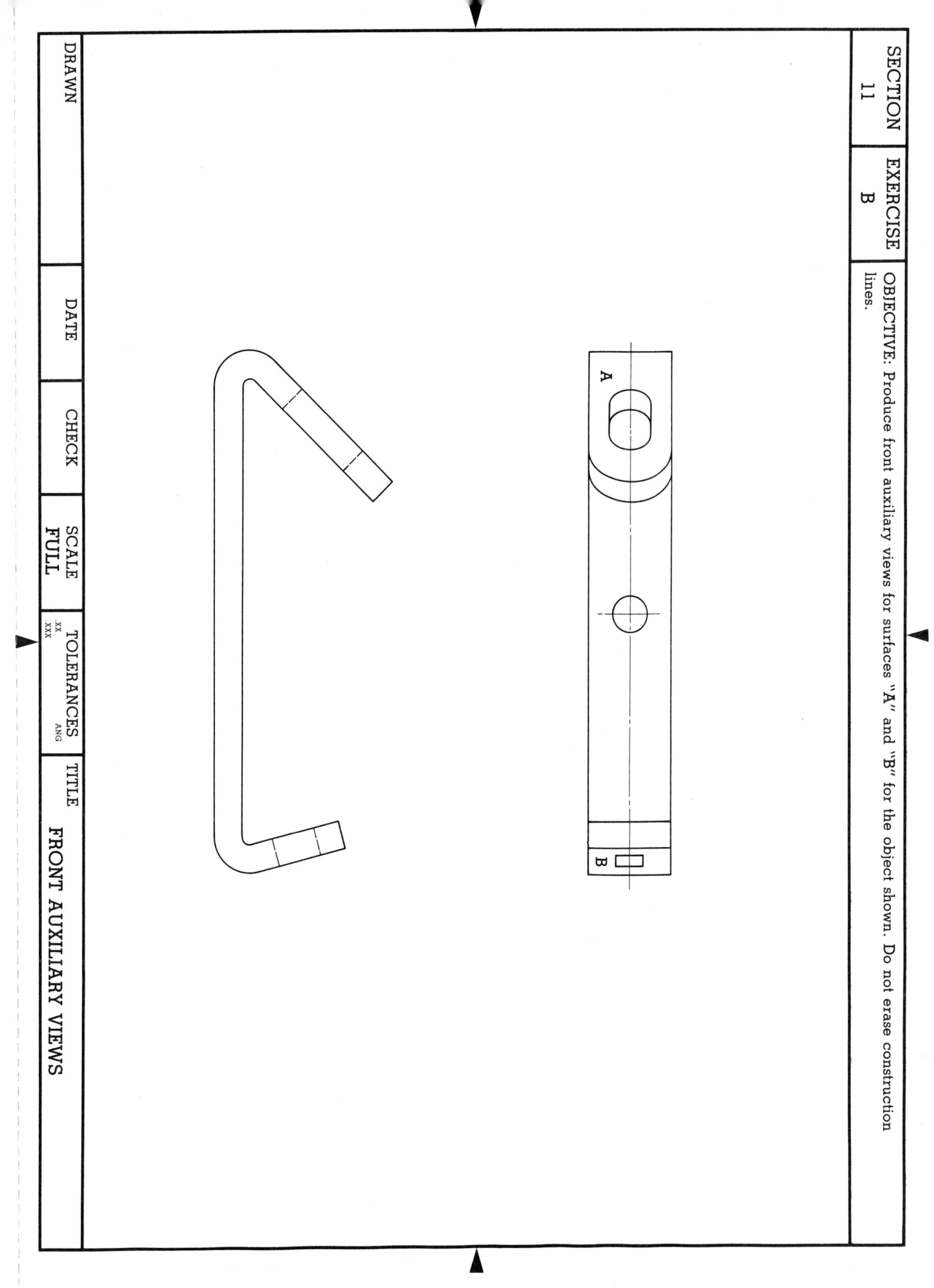

SECTION	EXERCISE	OBJECTIVE: Construct a top auxiliary view of the part shown. Do not erase construction lines.
11	C	

DRAWN	DATE	CHECK	SCALE FULL	TOLERANCES xx xxx ANG	TITLE TOP AUXILIARY VIEW

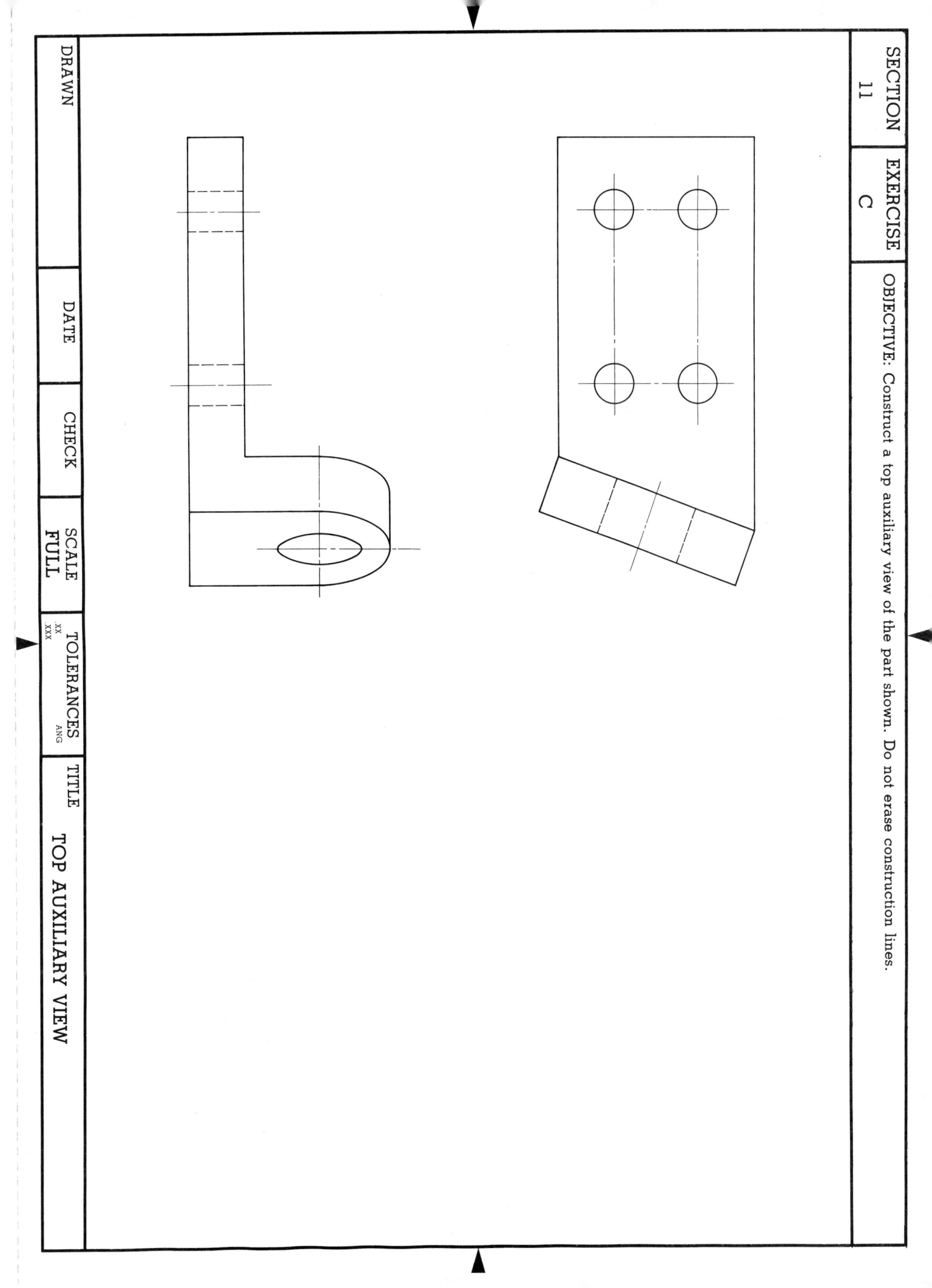

SECTION	EXERCISE	OBJECTIVE: Scale the given model and draw the front, top, and right side view. In addition, construct a side auxiliary view of slanted surface A. Do not dimension.
11	D	

FRONT VIEW

A

| DRAWN | DATE | CHECK | SCALE FULL | TOLERANCES XX XXX ANG | TITLE SIDE AUXILIARY VIEW |

SECTION	EXERCISE
11	E

OBJECTIVE: For the object illustrated below draw the primary (front) auxiliary view of surface "A" and a secondary auxiliary view showing surface "B". Show all construction lines.

| DRAWN | DATE | CHECK | SCALE FULL | TOLERANCES XX XXX ANG | TITLE PRIMARY AND SECONDARY AUX. VIEWS |

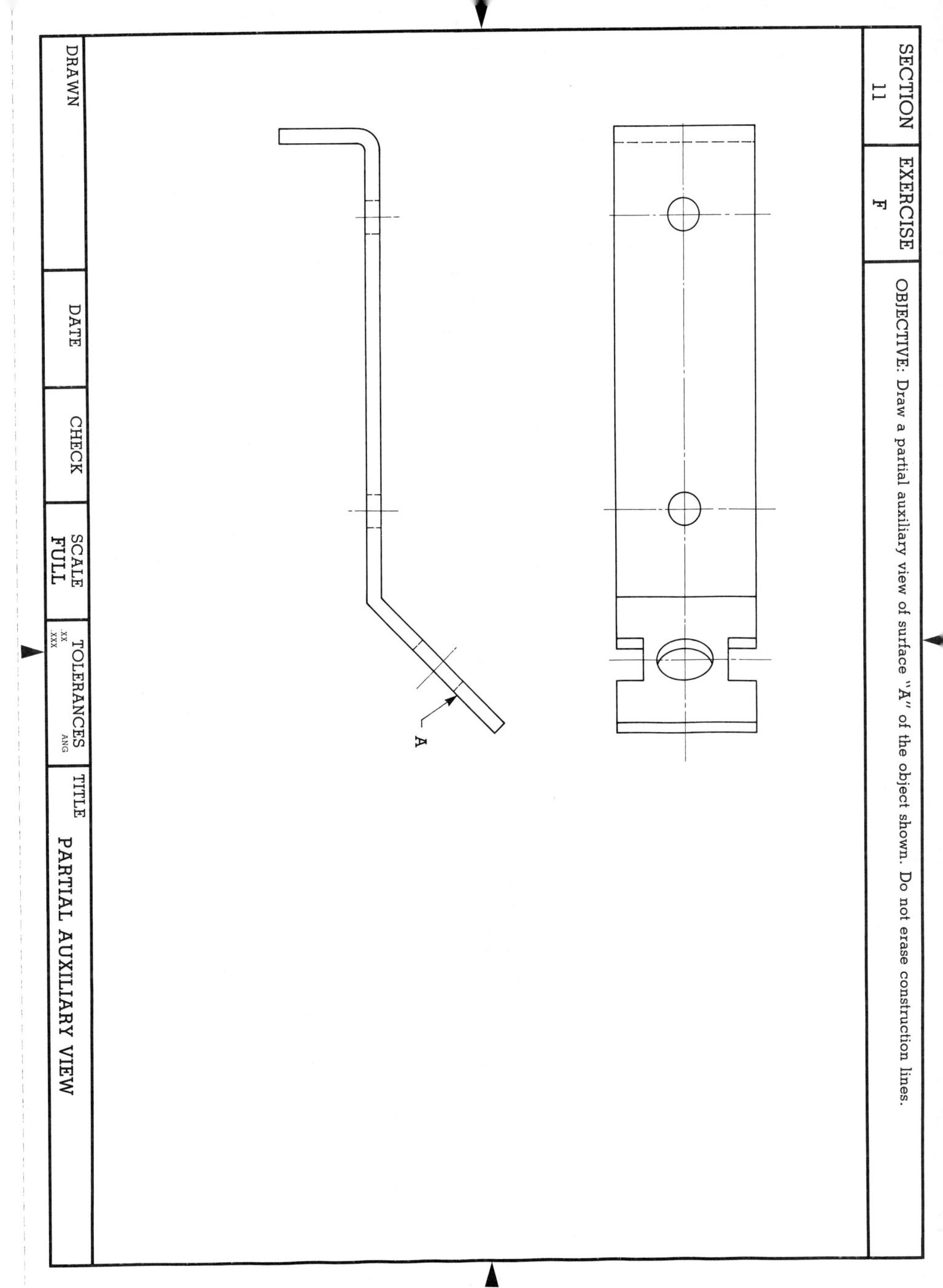

SECTION 11 EXERCISE F

OBJECTIVE: Draw a partial auxiliary view of surface "A" of the object shown. Do not erase construction lines.

TITLE: PARTIAL AUXILIARY VIEW

SCALE: FULL

SECTION	EXERCISE	OBJECTIVE: Complete the top view of the tube section shown. Construct curved surface auxiliary views for surfaces "A" and "B". Show all points, letters, intersections, and construction lines.
11	G	

DRAWN	DATE	CHECK	SCALE FULL	TOLERANCES XX XXX ANG	TITLE CURVED SURFACE AUXILIARY VIEW

| SECTION | EXERCISE | OBJECTIVE: Given the front view, left and right auxiliary views of the object shown, complete the top view, left side view, and right side views. |
| 11 | H | |

| DRAWN | DATE | CHECK | SCALE FULL | TOLERANCES XX XXX ANG | TITLE AUXILIARY VIEWS |

SECTION	EXERCISE	
12	A	OBJECTIVE: Given the thin-gage metal strap shown, draw two views of the object, then produce a development drawing. Neglect material thickness. Add all dimensions including developed length. Note: Dimensions are metric.

Dimensions shown on isometric: 50.8, 19, 25.4 TYP, 19 TYP, 15.9 TYP, 12.7

DRAWN	DATE	CHECK	SCALE FULL	TOLERANCES XX XXX ANG	TITLE
					DEVELOPMENT

SECTION	EXERCISE	OBJECTIVE: Draw two views of the given thin-gage metal bracket shown, then produce a development drawing. Neglect material thickness. Include all dimensions including the developed length.
12	B	

Dimensions shown on bracket: 1.25, .75, 1.38, 1.00, 2.00, 1.50

DRAWN	DATE	CHECK	SCALE FULL	TOLERANCES XX XXX ANG	TITLE DEVELOPMENT

SECTION	EXERCISE	
12	C	OBJECTIVE: Using the parallel line development method construct the development of the truncated prism shown. Include top and bottom surfaces. Use .19 wide tabs to join prism after cutting out the development.

DRAWN	DATE	CHECK	SCALE FULL	TOLERANCES XX XXX ANG	TITLE DEVELOPMENT - TRUNCATED PRISM

1.12
1.75
1.00
.88

| SECTION 12 | EXERCISE D | OBJECTIVE: Using the parallel line development method construct the development of the truncated prism shown. Include top and bottom surfaces. Use .19 wide tabs to join prism after cutting out the development. Dimensions are metric. |

30.2
50.8
31.8

| DRAWN | DATE | CHECK | SCALE FULL | TOLERANCES XX XXX ANG | TITLE DEVELOPMENT - TRUNCATED PRISM |

SECTION	EXERCISE	OBJECTIVE: Using the parallel line development method construct the development of the truncated cylinder shown. Include top and bottom surfaces. Use .19 wide tabs to join cylinder after cutting out the development.
12	E	

.94
1.62
Ø1.38

DRAWN	DATE	CHECK	SCALE FULL	TOLERANCES XX XXX ANG	TITLE DEVELOPMENT - TRUNCATED CYLINDER

SECTION	EXERCISE	
12	F	OBJECTIVE: Using the radial line development method construct the development of the truncated pyramid shown. Include top and bottom surfaces. Use .19 wide tabs to join pyramid after cutting out the development. Dimensions are metric.

DRAWN	DATE	CHECK	SCALE FULL	TOLERANCES XX / XXX / ANG	TITLE DEVELOPMENT - TRUNCATED PYRAMID

SECTION	EXERCISE	OBJECTIVE: Using the radial line development method construct the development of the truncated cone shown. Include top and bottom surfaces.
12	G	

DRAWN	DATE	CHECK	SCALE **HALF**	TOLERANCES XX XXX ANG	TITLE **DEVELOPMENT - TRUNCATED CONE**

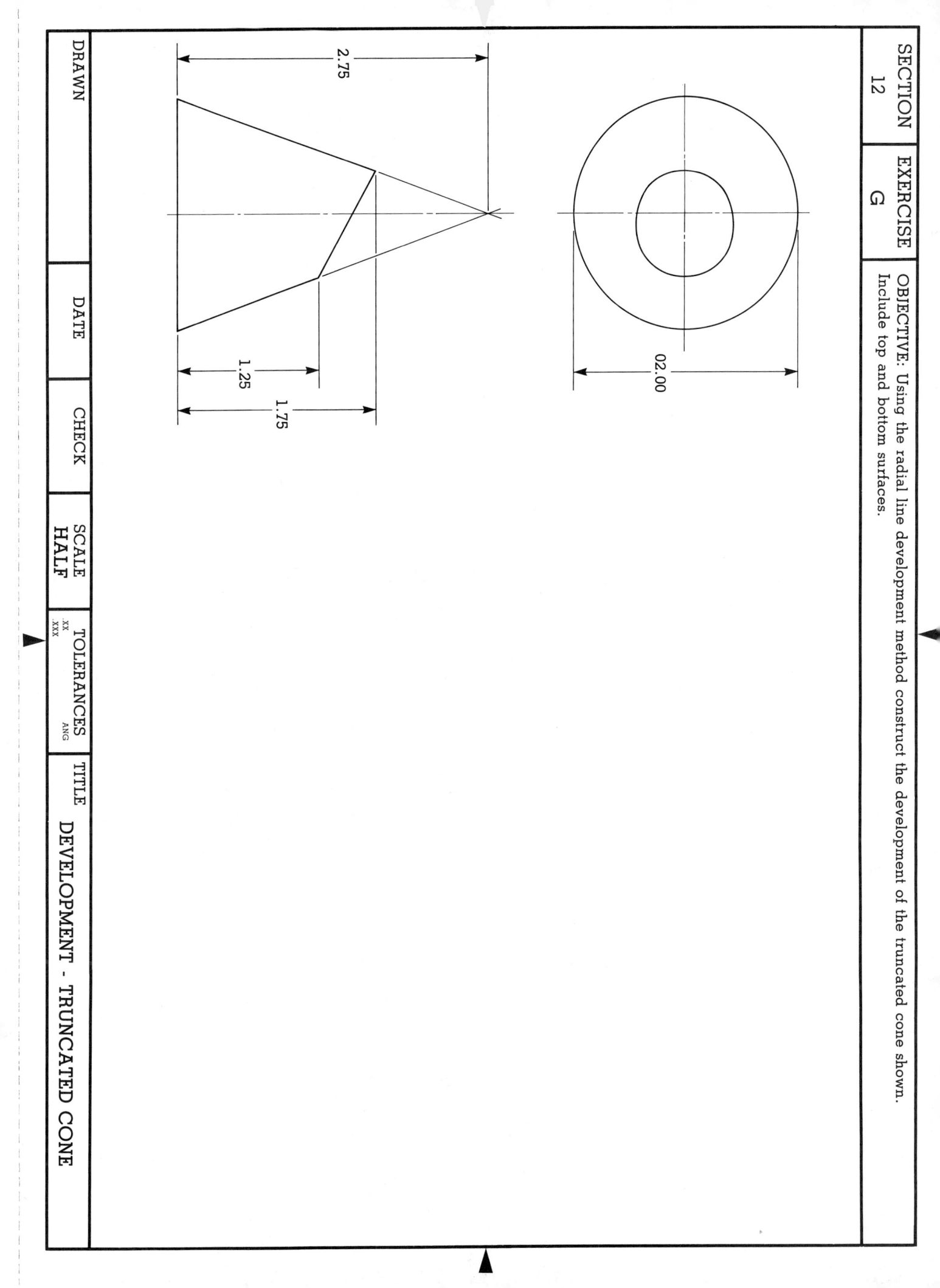

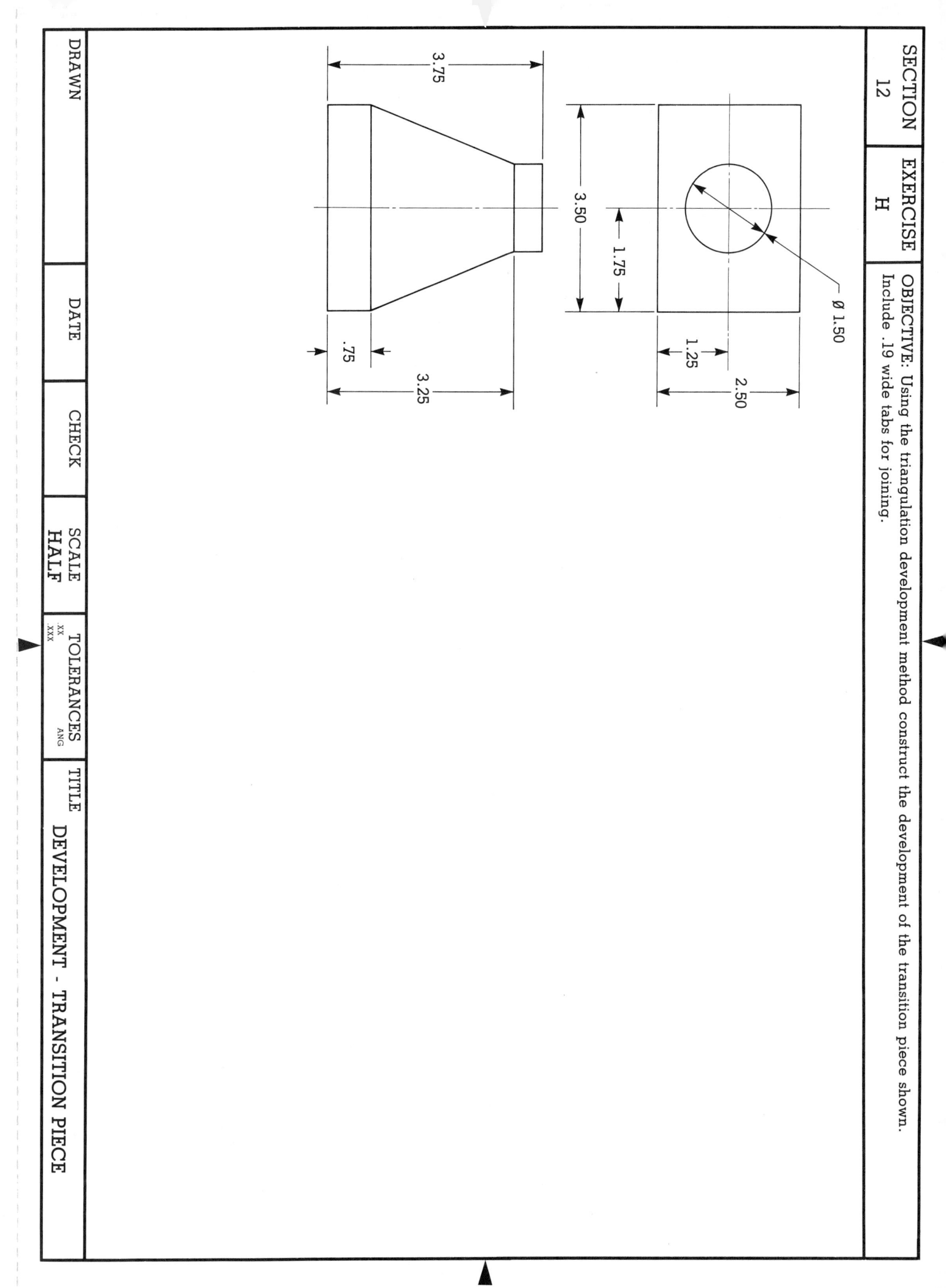

SECTION	EXERCISE	OBJECTIVE: Using the triangulation development method construct the development of the transition piece shown. Include .19 wide tabs for joining.
12	H	

DRAWN	DATE	CHECK	SCALE HALF	TOLERANCES XX XXX ANG	TITLE DEVELOPMENT - TRANSITION PIECE

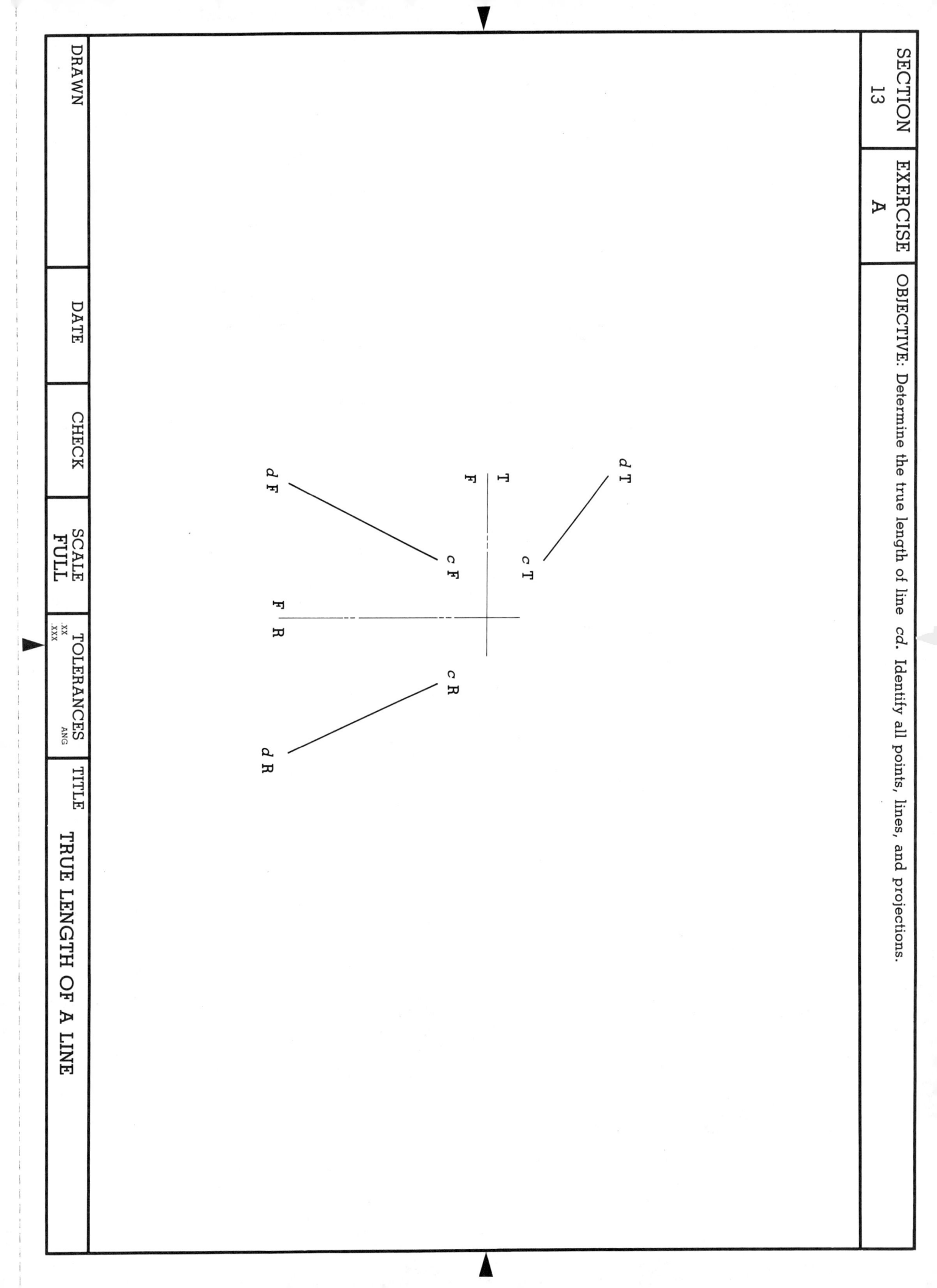

SECTION	EXERCISE	OBJECTIVE: Determine the true length of line cd. Identify all points, lines, and projections.
13	A	

Labels visible on drawing: d_T, T / F, c_T, d_F, c_F, F / R, c_R, d_R

| DRAWN | DATE | CHECK | SCALE FULL | TOLERANCES XX XXX ANG | TITLE TRUE LENGTH OF A LINE |

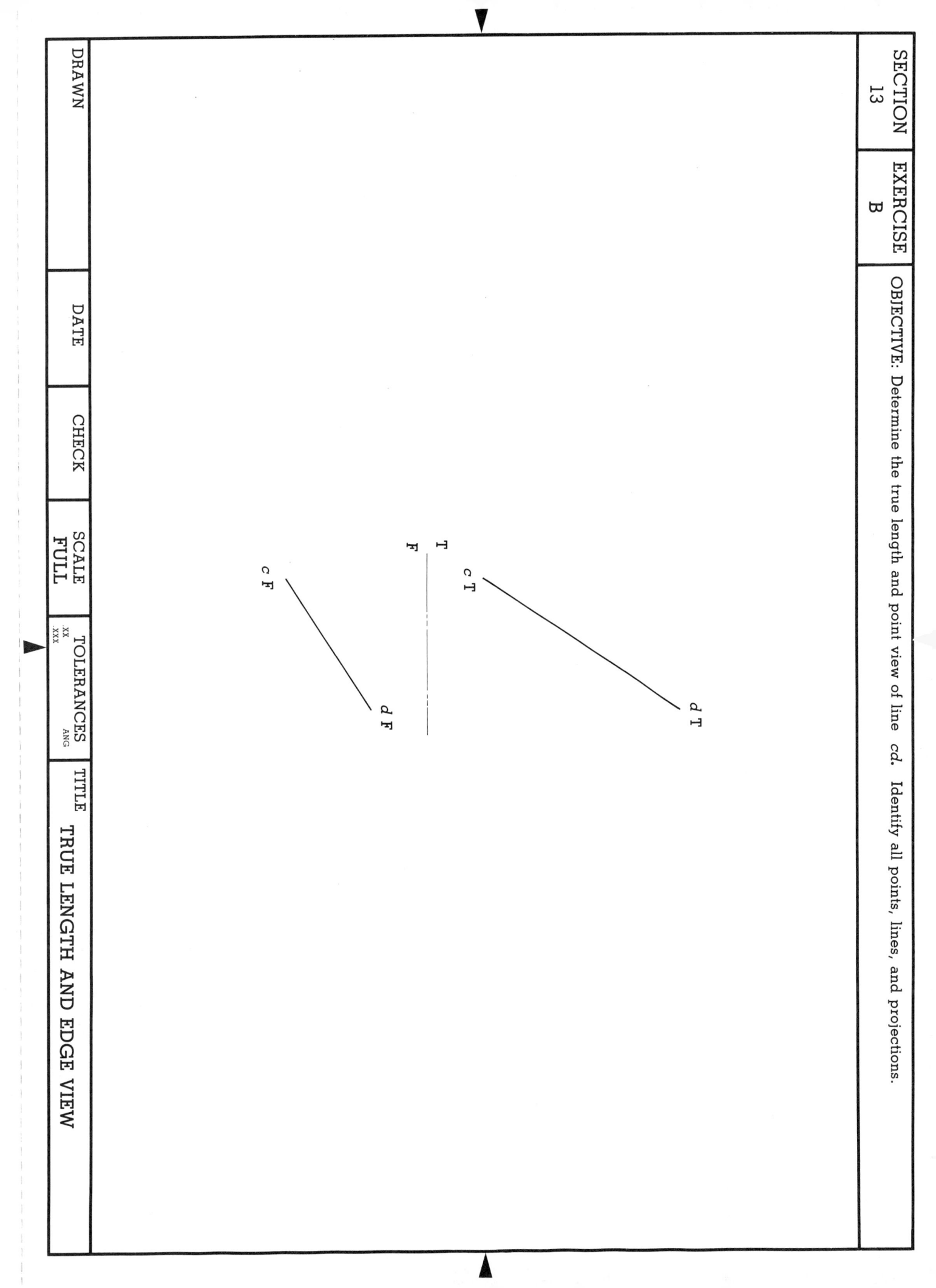

| SECTION | EXERCISE | OBJECTIVE: Determine the true length and point view of line cd. Identify all points, lines, and projections. |
| 13 | B | |

cF
dF
F|T
cT
dT

TITLE: TRUE LENGTH AND EDGE VIEW
SCALE: FULL
TOLERANCES .XX .XXX ANG
DRAWN
DATE
CHECK

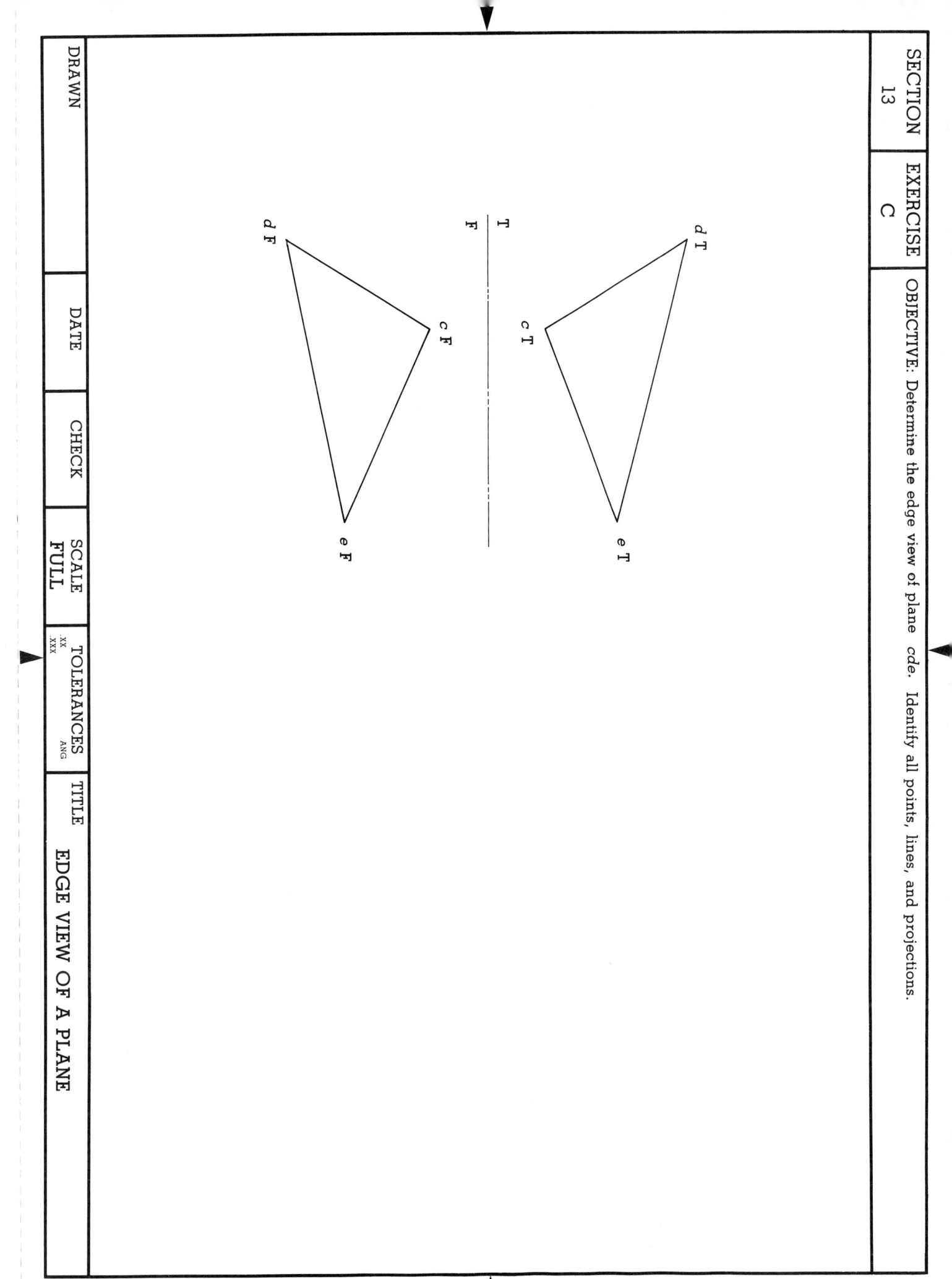

SECTION	EXERCISE	OBJECTIVE: Determine the edge view of plane *cde*. Identify all points, lines, and projections.
13	C	

d T
c T
e T

T
F

d F
c F
e F

DRAWN	DATE	CHECK	SCALE FULL	TOLERANCES XX XXX ANG	TITLE EDGE VIEW OF A PLANE

| SECTION 13 | EXERCISE D | OBJECTIVE: Determine the true size and shape of plane *cde*. Identify all points, lines, projections. |

dT

cT

eT

T

F

dF

cF

eF

| DRAWN | DATE | CHECK | SCALE FULL | TOLERANCES XX XXX ANG | TITLE **TRUE SIZE AND SHAPE** |

SECTION	EXERCISE	OBJECTIVE: Determine the true distance between plane *cde* and point *x* in space. Identify all points, lines, and projections.
13	E	

| DRAWN | DATE | CHECK | SCALE FULL | TOLERANCES XX / XXX / ANG | TITLE **TRUE DISTANCE BETWEEN A PLANE SURFACE AND A POINT** |

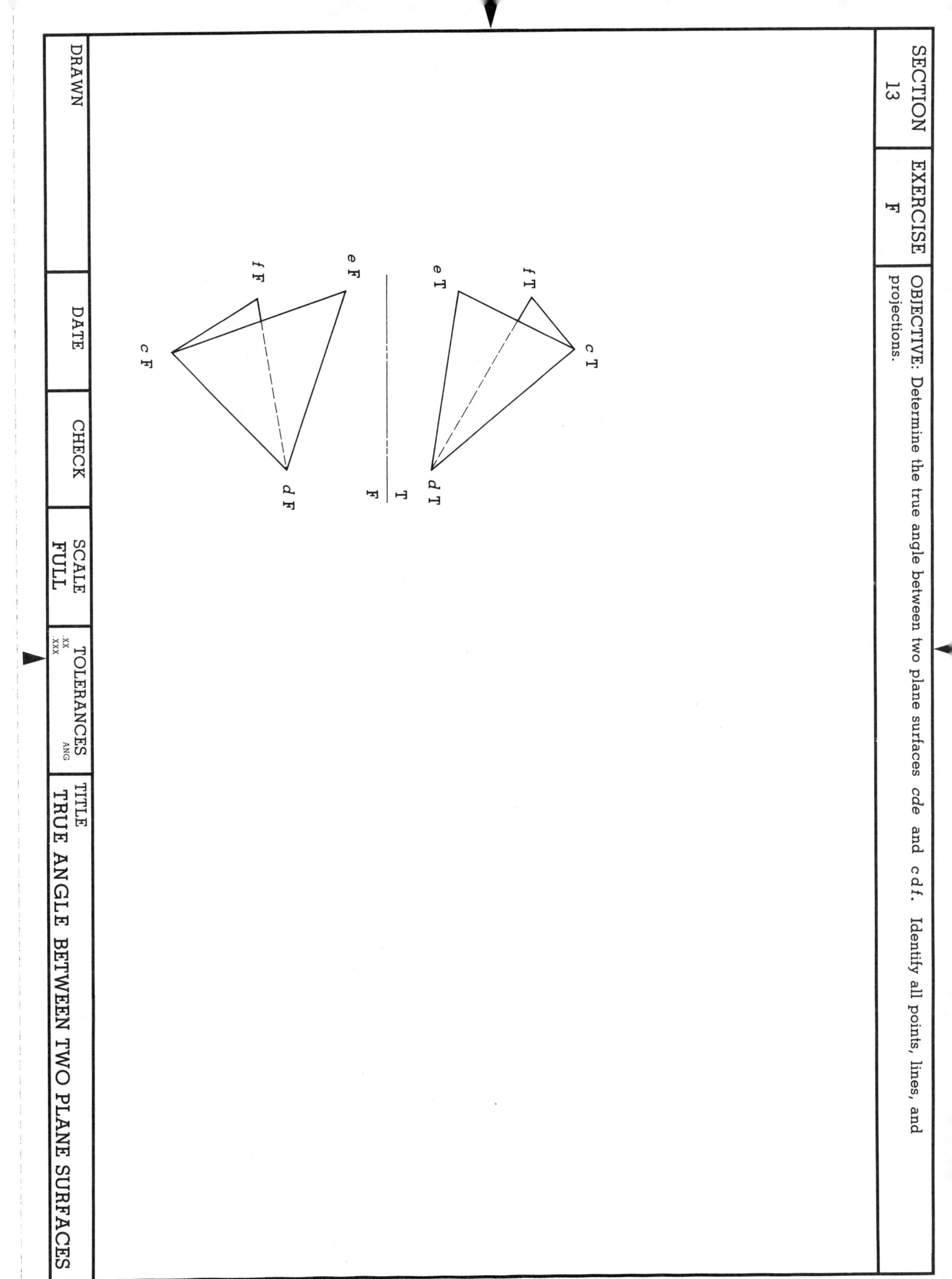

SECTION	EXERCISE	
13	F	OBJECTIVE: Determine the true angle between two plane surfaces cde and cdf. Identify all points, lines, and projections.

DRAWN	DATE	CHECK	SCALE FULL	TOLERANCES XX XXX ANG	TITLE TRUE ANGLE BETWEEN TWO PLANE SURFACES

SECTION	EXERCISE	OBJECTIVE: Determine the piercing point (intersection) of line *xy* through plane *abc*. Identify the piercing point in the front view of the plane.
13	G	

TITLE

INTERSECTION OF A LINE & A PLANE

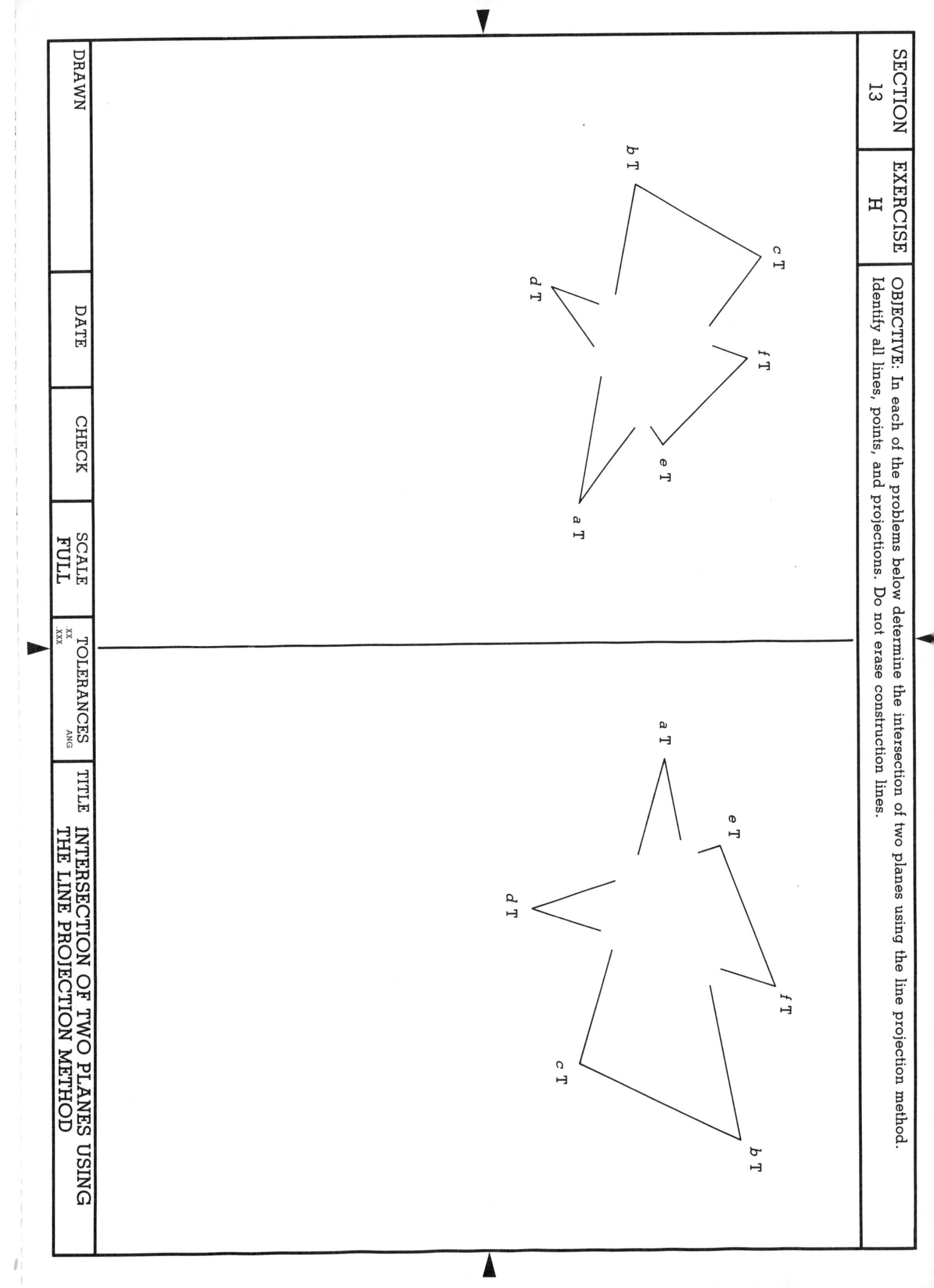

SECTION	EXERCISE	OBJECTIVE: In each of the problems below determine the intersection of two planes using the line projection method. Identify all lines, points, and projections. Do not erase construction lines.
13	H	

| DRAWN | DATE | CHECK | SCALE FULL | TOLERANCES .XX .XXX ANG | TITLE **INTERSECTION OF TWO PLANES USING THE LINE PROJECTION METHOD** |

SECTION	EXERCISE	OBJECTIVE: Produce an isometric pictorial instrument drawing of each of the objects shown. Measurements may be transferred with a scale or dividers.
14	A	

DRAWN	DATE	CHECK	SCALE FULL	TOLERANCES XX / XXX / ANG	TITLE
					PICTORIAL, ISOMETRIC

SECTION	EXERCISE
14	B

OBJECTIVE: Each of the objects shown consists of isometric and non-isometric lines. Produce an isometric pictorial instrument drawing of each. Transfer measurements with a scale or dividers.

DRAWN		DATE	CHECK	SCALE FULL	TOLERANCES XX XXX ANG	TITLE
						PICTORIAL, ISOMETRIC

SECTION	EXERCISE	OBJECTIVE: Produce an isometric pictorial instrument drawing of the curved surfaced objects below. Transfer measurements.
14	C	

DRAWN	DATE	CHECK	SCALE FULL	TOLERANCES XX XXX ANG	TITLE
					PICTORIAL, ISOMETRIC

SECTION	14
EXERCISE	D

OBJECTIVE: Draw an oblique cabinet pictorial instrument drawing of the object shown. Center in the space provided.

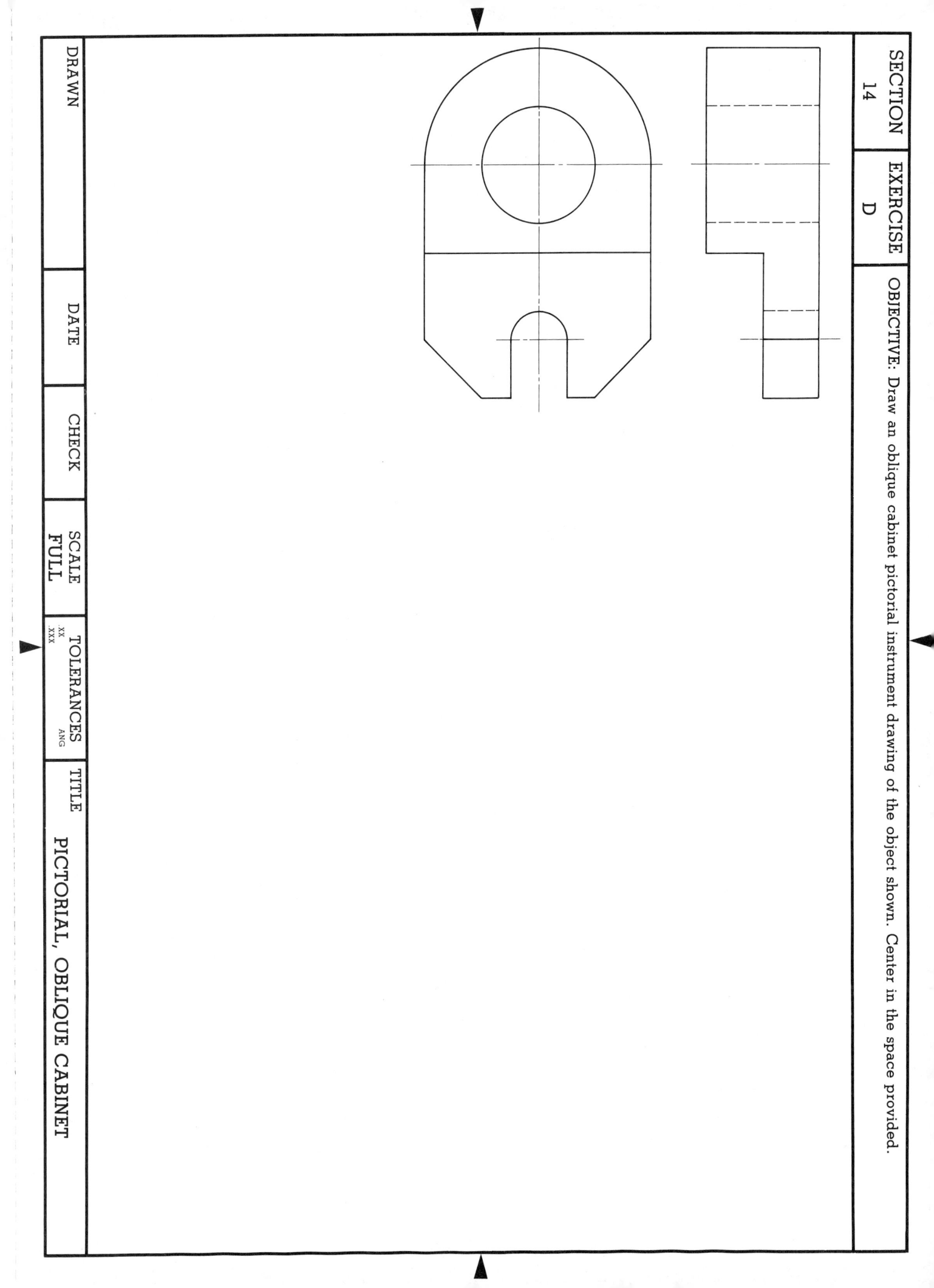

DRAWN		
DATE	CHECK	SCALE **FULL**

TOLERANCES	XX
ANG	XXX

TITLE

PICTORIAL, OBLIQUE CABINET

SECTION	EXERCISE	OBJECTIVE: Produce a twice size oblique cabinet pictorial instrument drawing of the object shown in the space provided.
14	E	

DRAWN	DATE	CHECK	SCALE 2:1	TOLERANCES XX XXX ANG	TITLE
					PICTORIAL, OBLIQUE CABINET

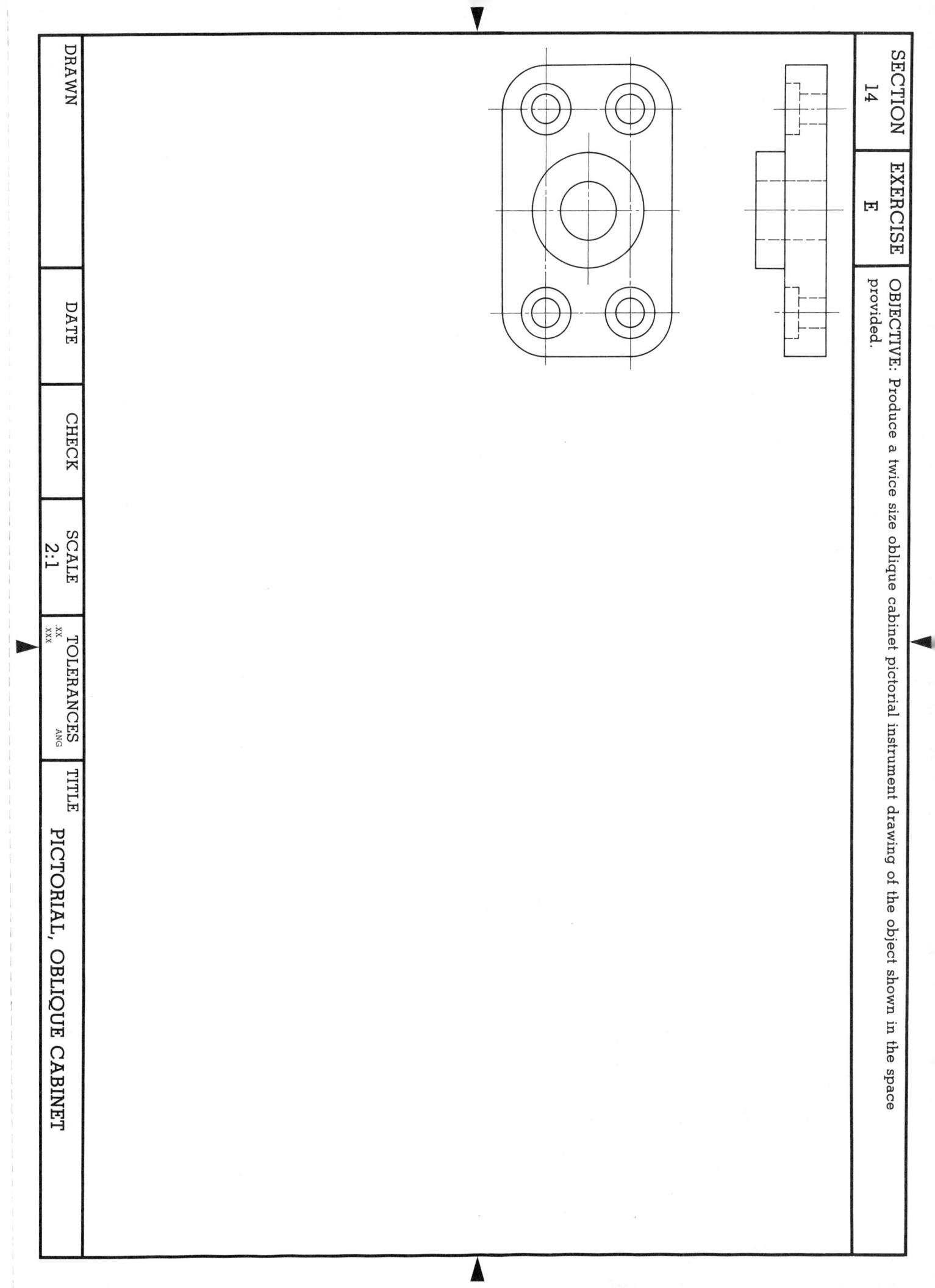

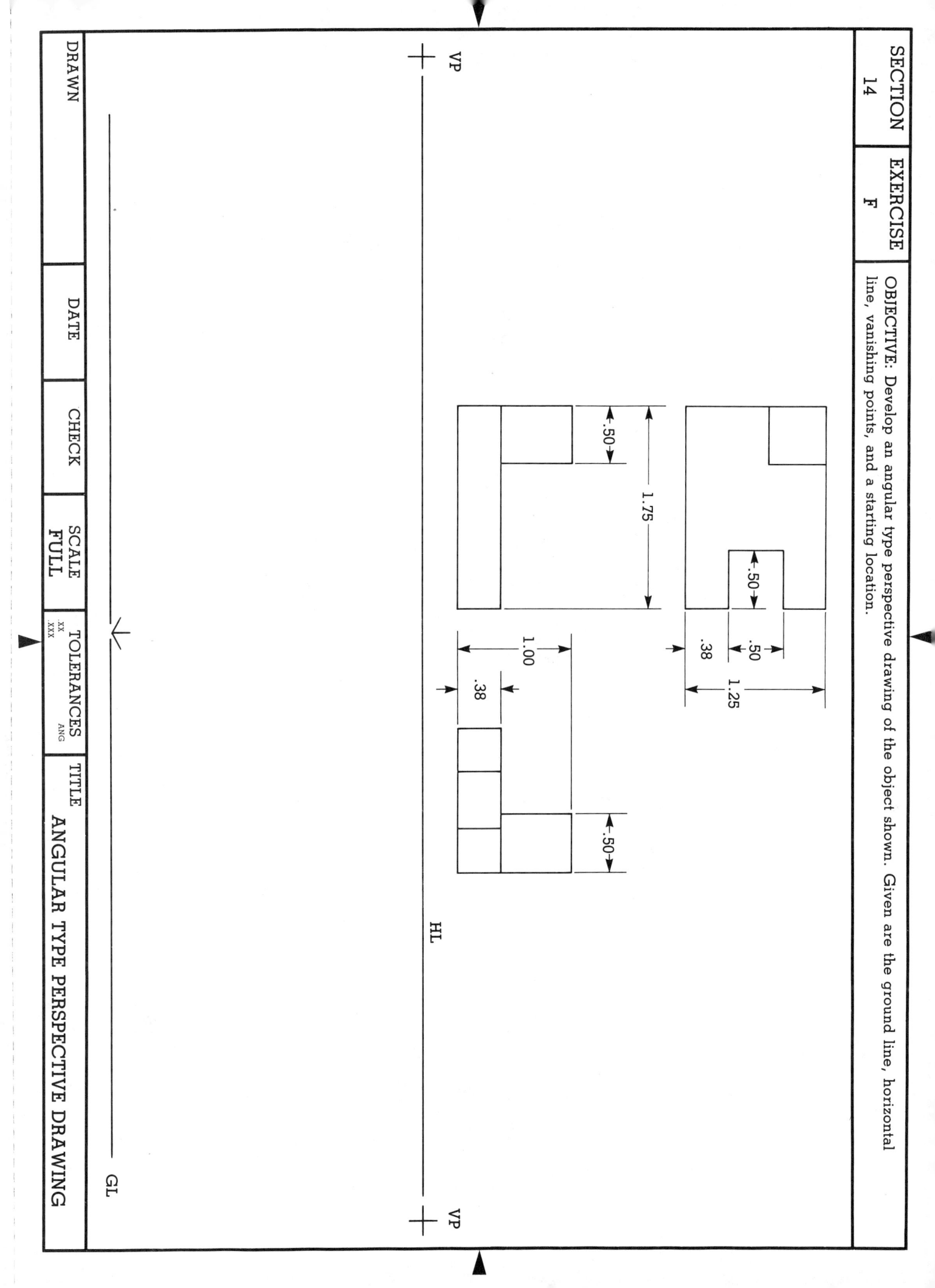

SECTION	EXERCISE		
14	F	OBJECTIVE: Develop an angular type perspective drawing of the object shown. Given are the ground line, horizontal line, vanishing points, and a starting location.	

VP

HL

VP   VP

GL

DRAWN	DATE	CHECK	SCALE FULL	TOLERANCES XX XXX ANG	TITLE
					ANGULAR TYPE PERSPECTIVE DRAWING

SECTION	EXERCISE		
14	G		

OBJECTIVE: Given the object below (top and right side view), picture plane, station point, and ground line. Develop an angular (two-vanishing point) type perspective drawing of the object.

DRAWN	DATE	CHECK	SCALE **FULL**	TOLERANCES XX XXX ANG	TITLE **ANGULAR TYPE PERSPECTIVE DRAWING**

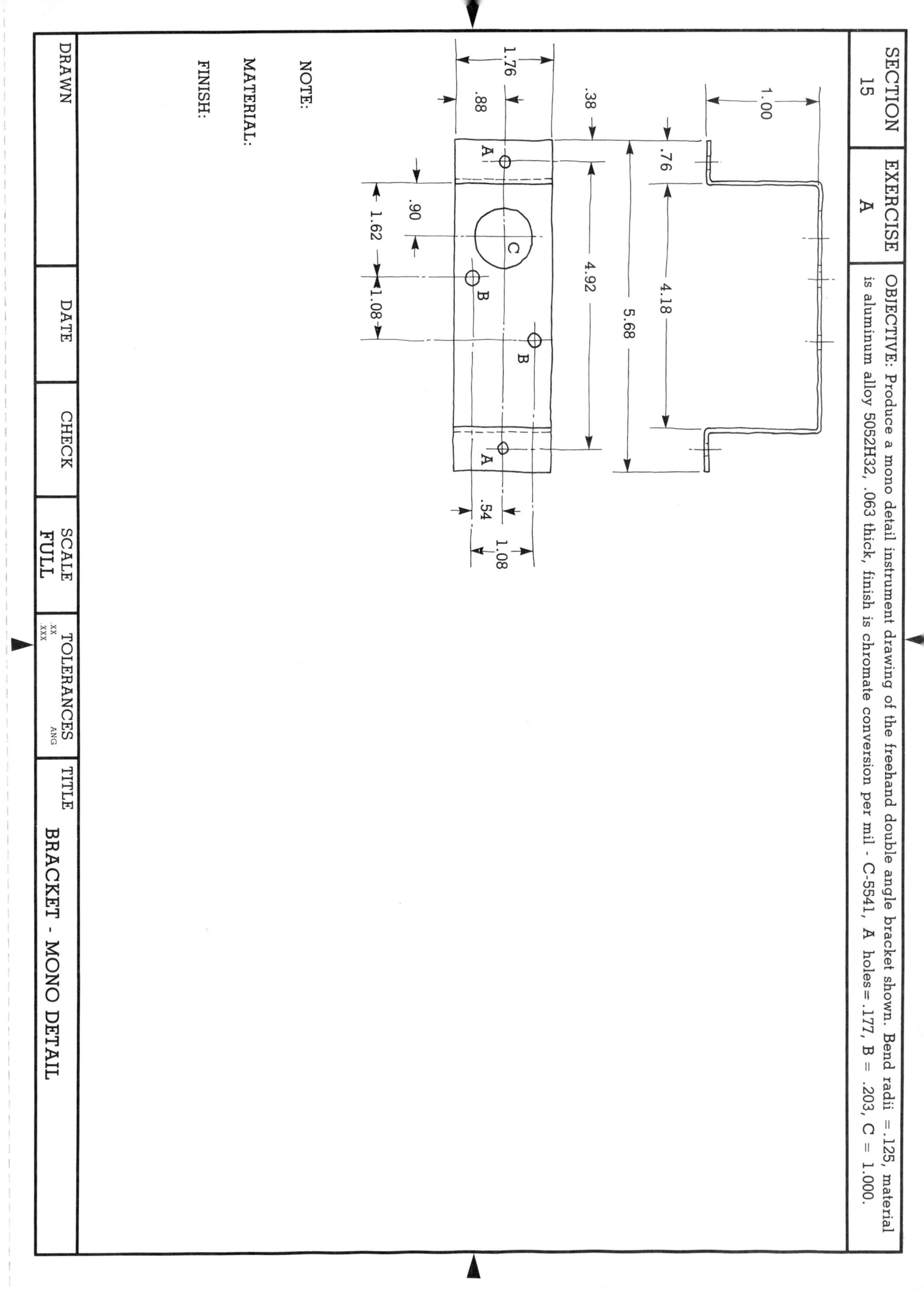

SECTION	EXERCISE	
15	A	OBJECTIVE: Produce a mono detail instrument drawing of the freehand double angle bracket shown. Bend radii = .125, material is aluminum alloy 5052H32, .063 thick, finish is chromate conversion per mil - C-5541, A holes = .177, B = .203, C = 1.000.

NOTE:

MATERIAL:

FINISH:

DRAWN	DATE	CHECK	SCALE **FULL**	TOLERANCES .xx .xxx ANG	TITLE **BRACKET - MONO DETAIL**

SECTION	EXERCISE	
15	B	OBJECTIVE: Produce a multi detail drawing of the freehand plumb bob shown. Specifications are identified below.

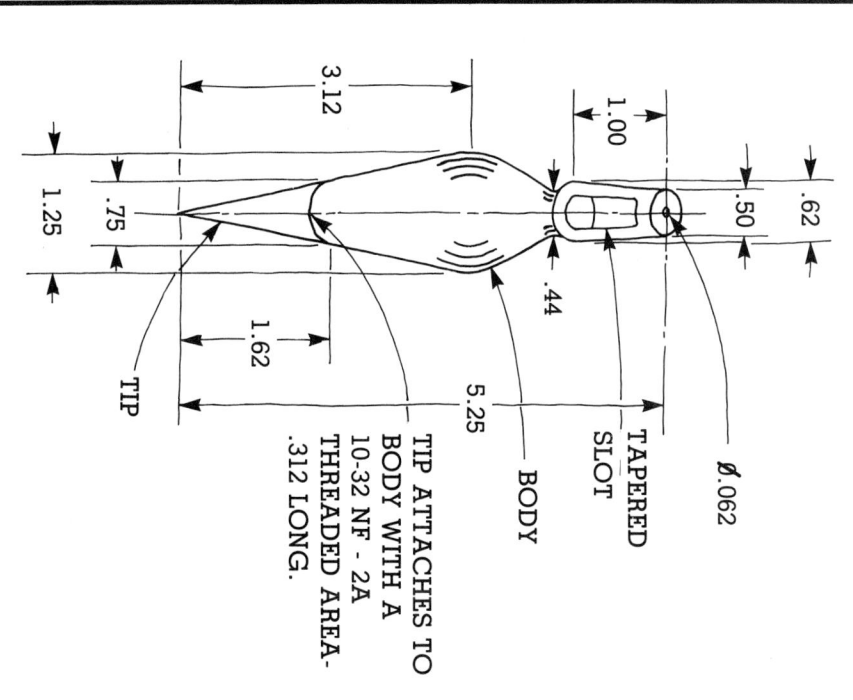

BODY:
MATERIAL: DIE CAST
 ALUMINUM ALLOY
FINISH: BRIGHT ORANGE
 ENAMEL - SEMI-GLOSS

TIP:
MATERIAL: 6061T6
 ALUMINUM ROD
FINISH: CLEAR ANODIZE

Dimensions shown: 3.12, 1.00, .62, .50, Ø.062, 1.25, .75, .44, 1.62, 5.25

Labels: TIP, BODY, TAPERED SLOT, TIP ATTACHES TO BODY WITH A 10-32 NF - 2A THREADED AREA - .312 LONG.

DRAWN	DATE	CHECK	SCALE FULL	TOLERANCES XX XXX ANG	TITLE PLUMB BOB - MULTI DETAIL

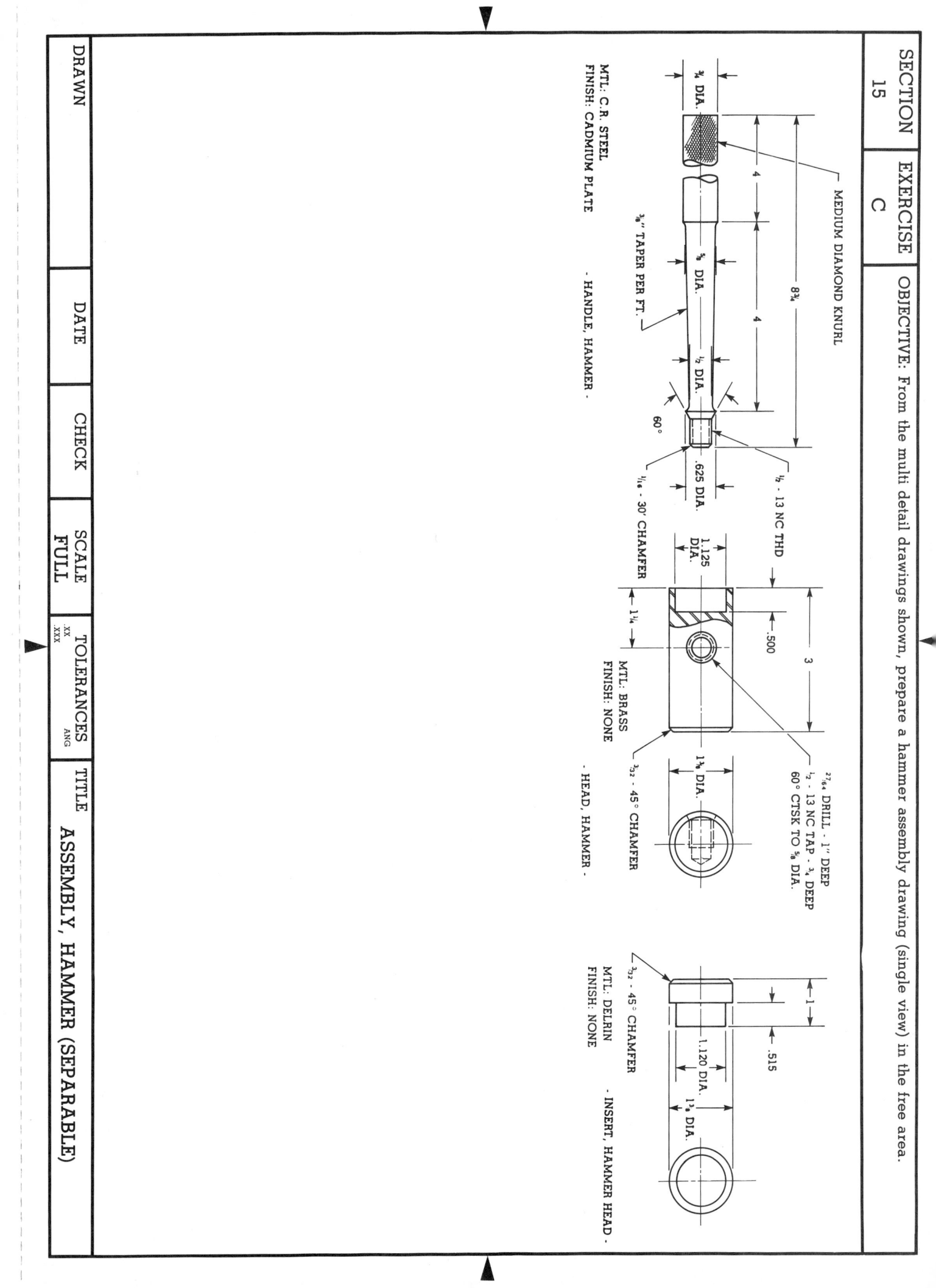

SECTION	EXERCISE	OBJECTIVE: From the multi detail drawings shown, prepare a hammer assembly drawing (single view) in the free area.
15	C	

MEDIUM DIAMOND KNURL

¾ DIA.
8¾
4
4
⅝ DIA.
½ DIA.
⅜" TAPER PER FT.
60°
.625 DIA.
1/16 - 30° CHAMFER
½ - 13 NC THD

MTL: C.R. STEEL
FINISH: CADMIUM PLATE

- HANDLE, HAMMER -

1.125 DIA.
1¼
.500
3
1¾ DIA.
3/32 - 45° CHAMFER

27/64 DRILL - 1" DEEP
½ - 13 NC TAP - ¾ DEEP
60° CTSK TO ⅝ DIA.

MTL: BRASS
FINISH: NONE

- HEAD, HAMMER -

1
.515
1.120 DIA.
1⅜ DIA.
3/32 - 45° CHAMFER

MTL: DELRIN
FINISH: NONE

- INSERT, HAMMER HEAD -

DRAWN	DATE	CHECK	SCALE FULL	TOLERANCES XX XXX ANG	TITLE ASSEMBLY, HAMMER (SEPARABLE)

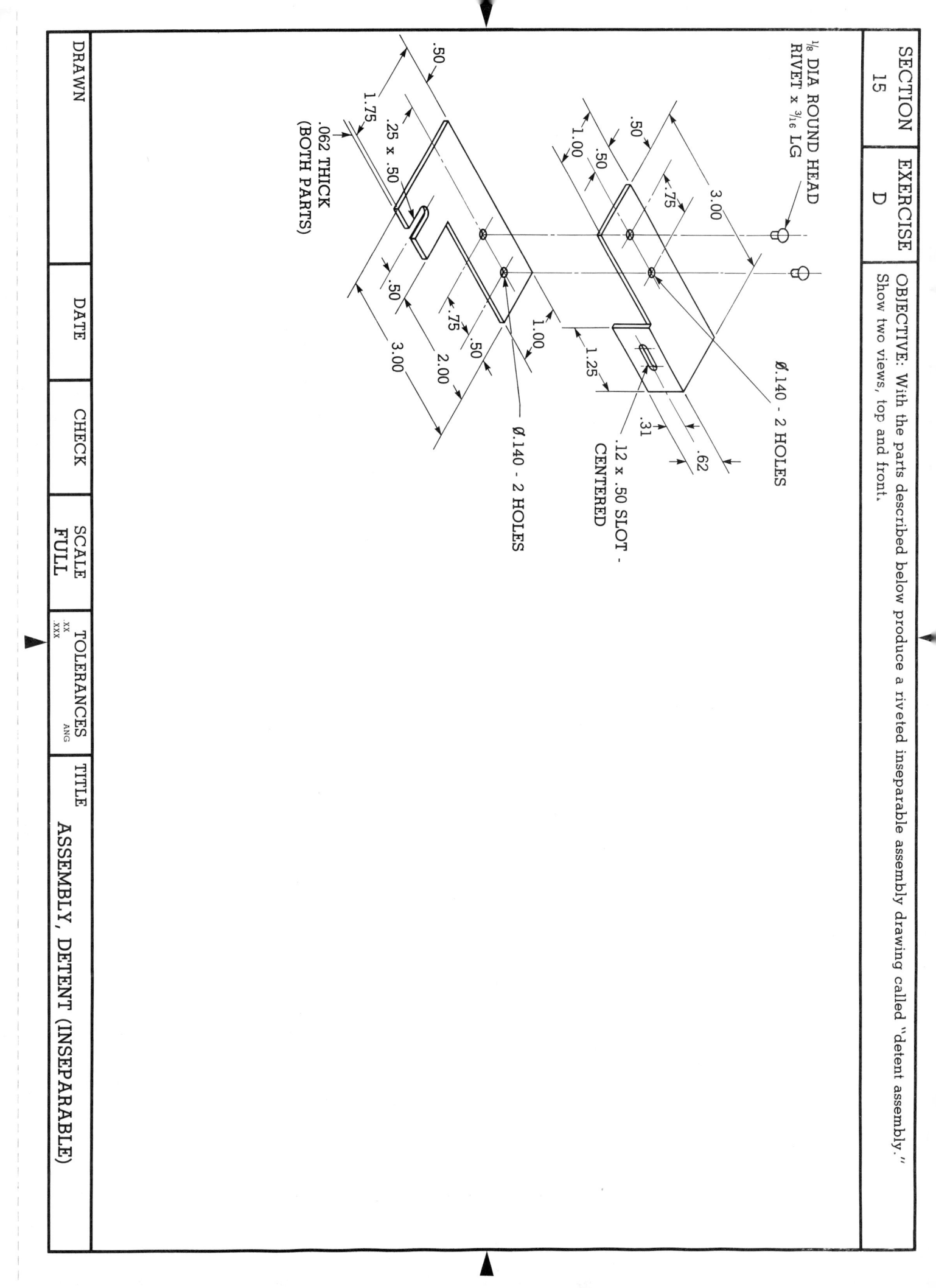

SECTION	EXERCISE	OBJECTIVE: With the parts described below produce a riveted inseparable assembly drawing called "detent assembly." Show two views, top and front.
15	D	

⌀.140 - 2 HOLES

1/8 DIA ROUND HEAD RIVET x 3/16 LG

.50
.50
1.00
.75
3.00
1.25
.31
.62
.12 x .50 SLOT - CENTERED
1.00
.50
.75
2.00
3.00
⌀.140 - 2 HOLES
.50
.25 x .50
1.75
.062 THICK (BOTH PARTS)

| DRAWN | DATE | CHECK | SCALE FULL | TOLERANCES XX XXX ANG | TITLE ASSEMBLY, DETENT (INSEPARABLE) |

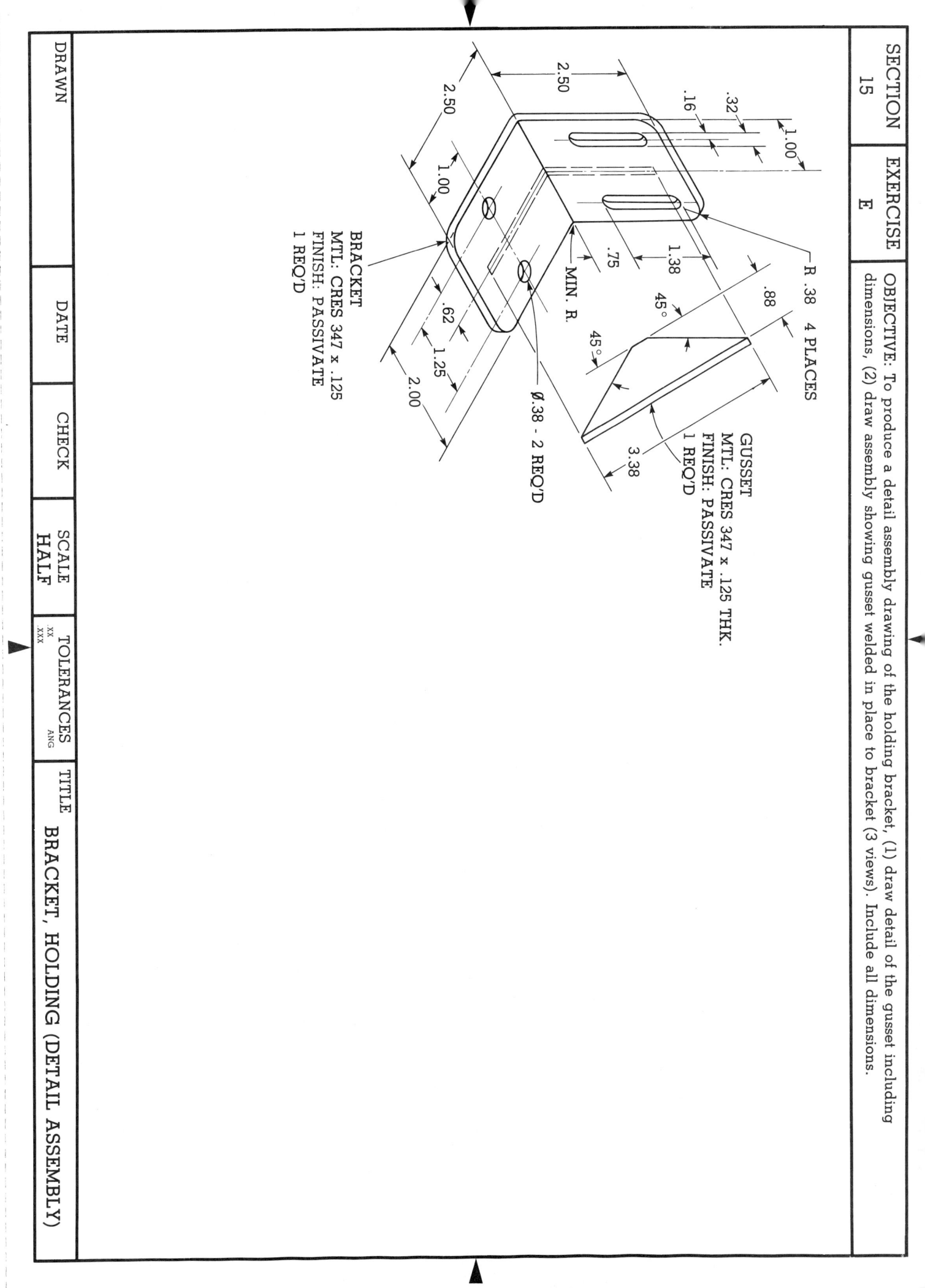

SECTION	EXERCISE	OBJECTIVE: To produce a detail assembly drawing of the holding bracket, (1) draw detail of the gusset including dimensions, (2) draw assembly showing gusset welded in place to bracket (3 views). Include all dimensions.
15	E	

R .38 4 PLACES

BRACKET
MTL: CRES 347 x .125
FINISH: PASSIVATE
1 REQ'D

GUSSET
MTL: CRES 347 x .125 THK.
FINISH: PASSIVATE
1 REQ'D

Ø.38 - 2 REQ'D

MIN. R.

DRAWN	DATE	CHECK	SCALE HALF	TOLERANCES .XX .XXX ANG	TITLE BRACKET, HOLDING (DETAIL ASSEMBLY)

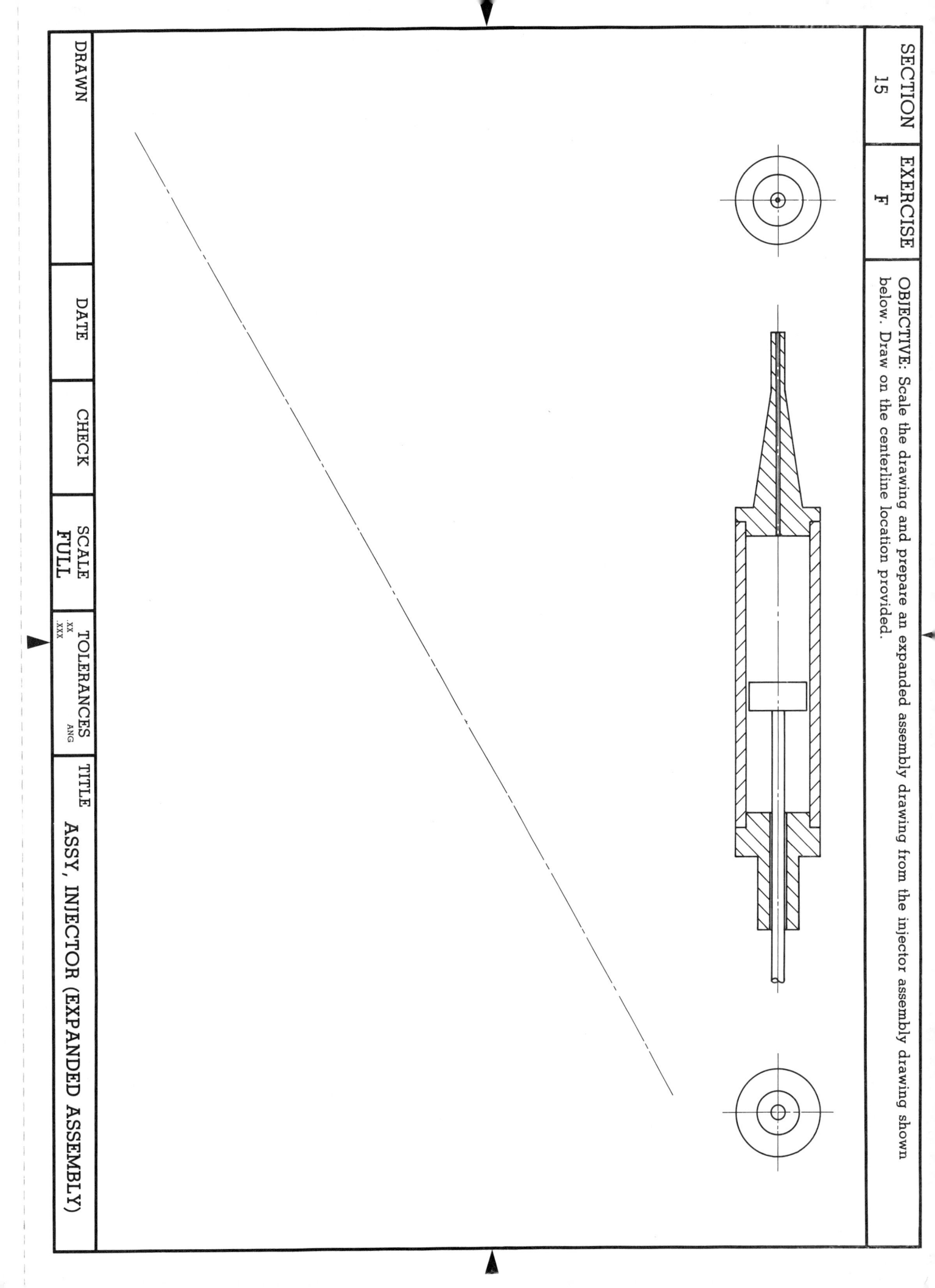

SECTION	EXERCISE	
15	F	OBJECTIVE: Scale the drawing and prepare an expanded assembly drawing from the injector assembly drawing shown below. Draw on the centerline location provided.

DRAWN	DATE	CHECK	SCALE FULL	TOLERANCES XX XXX ANG	TITLE ASSY, INJECTOR (EXPANDED ASSEMBLY)

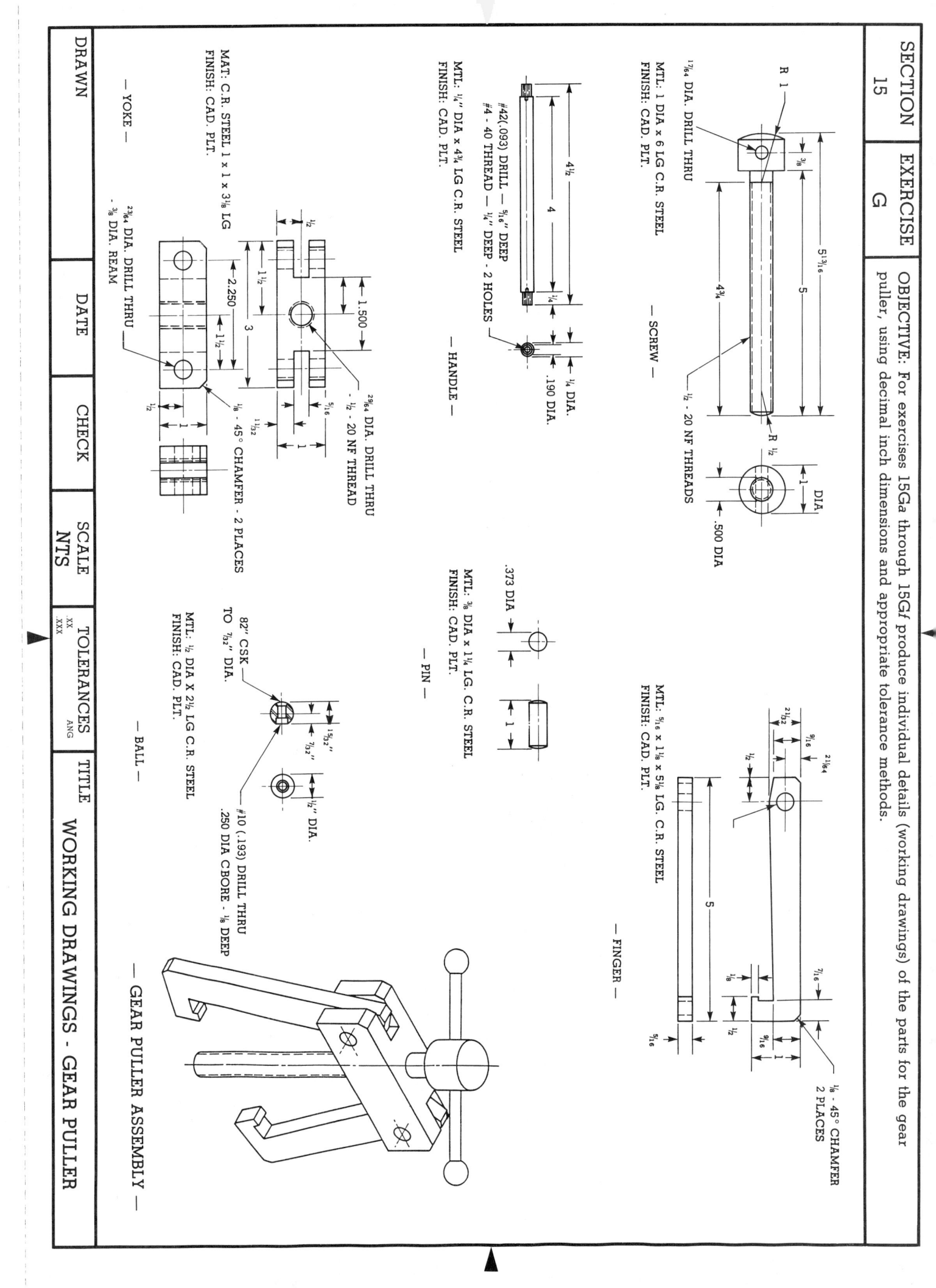

SECTION	EXERCISE	
15	G	OBJECTIVE: For exercises 15Ga through 15Gf produce individual details (working drawings) of the parts for the gear puller, using decimal inch dimensions and appropriate tolerance methods.

— SCREW —
MTL: 1 DIA × 6 LG C.R. STEEL
FINISH: CAD. PLT.
1⁷⁄₆₄ DIA. DRILL THRU

— HANDLE —
MTL: ¼ DIA × 4¾ LG C.R. STEEL
FINISH: CAD. PLT.
#42(.093) DRILL — ⁵⁄₁₆" DEEP
#4 - 40 THREAD — ¼" DEEP - 2 HOLES

— PIN —
MTL: ⅜ DIA × 1¼ LG. C.R. STEEL
FINISH: CAD. PLT.

— FINGER —
MTL: ⁵⁄₁₆ × 1⅛ × 5⅛ LG. C.R. STEEL
FINISH: CAD. PLT.
⅛ - 45° CHAMFER 2 PLACES

— YOKE —
MAT: C.R. STEEL 1 × 1 × 3⅛ LG
FINISH: CAD. PLT.
²³⁄₆₄ DIA. DRILL THRU
⅜ DIA. REAM
²⁹⁄₆₄ DIA. DRILL THRU
½ - 20 NF THREAD
⅛ - 45° CHAMFER - 2 PLACES

— BALL —
MTL: ½ DIA × 2½ LG C.R. STEEL
FINISH: CAD. PLT.
82° CSK TO ⁷⁄₃₂" DIA.
#10 (.193) DRILL THRU
.250 DIA CBORE - ⅛ DEEP

— GEAR PULLER ASSEMBLY —

DRAWN	DATE	CHECK	SCALE NTS	TOLERANCES XX / XXX / ANG	TITLE WORKING DRAWINGS - GEAR PULLER

SECTION	EXERCISE	OBJECTIVE: Produce a detail drawing of the gear puller ball (Exercise 15G). Include sufficient information for the part to be produced.
15	Ga	

DRAWN	DATE	CHECK	SCALE 2:1	TOLERANCES .XX .XXX ANG	TITLE
					BALL, GEAR PULLER

SECTION	EXERCISE	OBJECTIVE: Prepare a detail drawing of the gear puller screw (Exercise 15G). Include sufficient information for the part to be produced.
15	G*b*	

| DRAWN | DATE | CHECK | SCALE 1:1 | TOLERANCES XX XXX ANG | TITLE SCREW GEAR PULLER |

SECTION	EXERCISE	OBJECTIVE: Prepare a detail drawing of the gear puller pin (Exercise 15G). Include sufficient information to produce the part.
15	Gc	

DRAWN	DATE	CHECK	SCALE 2:1	TOLERANCES XX XXX ANG	TITLE PIN, GEAR PULLER

SECTION	EXERCISE	OBJECTIVE: Produce a detail drawing of the gear puller finger (Exercise 15G). Provide sufficient information to produce the part.
15	Gd	

DRAWN	DATE	CHECK	SCALE 1:1	TOLERANCES XX XXX ANG	TITLE FINGER, GEAR PULLER

SECTION	EXERCISE	OBJECTIVE: Prepare a detail drawing of the gear puller handle (Exercise 15G). Include sufficient information to produce the part.
15	Ge	

DRAWN	DATE	CHECK	SCALE 1:1	TOLERANCES XX XXX ANG	TITLE
					HANDLE, GEAR PULLER

SECTION	EXERCISE	
15	Gf	OBJECTIVE: Produce a detail drawing of the gear puller yoke (Exercise 15G). Provide sufficient information for the part to be produced.

DRAWN	DATE	CHECK	SCALE 1:1	TOLERANCES XX XXX ANG	TITLE
					YOKE, GEAR PULLER

SECTION	EXERCISE	
15	H	OBJECTIVE: Prepare an assembly drawing for the gear puller showing all parts and developing a parts list including parts identification. Use completed exercises 15Ga through 15Gf for reference.

DRAWN	DATE	CHECK	SCALE FULL	TOLERANCES XX XXX ANG	TITLE ASSEMBLY, GEAR PULLER

SECTION	EXERCISE	
15	1	OBJECTIVE: Produce a sheet metal detail drawing of the mounting bracket shown. Calculate the developed length using Appendix K for reference.

INSIDE BENDS = R .09
MATERIAL THICKNESS = .06

DEVELOPED LENGTH =

DRAWN	DATE	CHECK	SCALE FULL	TOLERANCES .XX .XXX ANG	TITLE BRACKET, MOUNTING

SECTION 15	EXERCISE J				
DRAWN	DATE	CHECK	SCALE FULL	TOLERANCES XX XXX ANG	TITLE BRACKET, OFFSET

OBJECTIVE: From the formed sheet metal detail drawing shown, prepare a blank (in the flat) drawing. Include all dimensions, holes, and slot. Calculate the developed length. Note: Dimensions are metric.

R 3.2 TYP.

Ø3.96 - 3 HOLES

.081
9.6
6.4
12.7
25.4
9.6
19
69.8
50.8
108
19
6.4
12.7
25.4
38.1
25.4
19.8

SECTION 16
PRACTICE EXERCISES

REALIZING THAT THE STUDENT MAY POSSESS VARYING DEGREES OF CAD PROFICIENCY, IT IS RECOMMENDED THAT A BASIC KNOWLEDGE OF CAD AND CAD SYSTEMS BE A PREREQUISITE FOR WORKING ON THE FOLLOWING EXERCISES.

THE FOLLOWING EXERCISES ARE TO BE COMPLETED USING A COMPUTER-AIDED DRAFTING (CAD) SYSTEM INCLUDING ITS ASSOCIATED SOFTWARE. MAINFRAME, MINI, OR MICROCOMPUTERS MAY BE USED TO DEVELOP THESE DRAWINGS. IN ADDITION, AN APPROPRIATE PLOTTING DEVICE TO PRODUCE A HARD COPY OF EACH EXERCISE IS RECOMMENDED.

16 A. OBJECTIVE: Produce a CAD drawing of the "A" size (8½ x 11 inch) and "B" size (11 x 17 inch) formats shown below. These formats are to be used as the border for all the exercises in this section. Add text/font information as shown using a font of your choice. Store formats in memory so that they may be "called up" when needed.

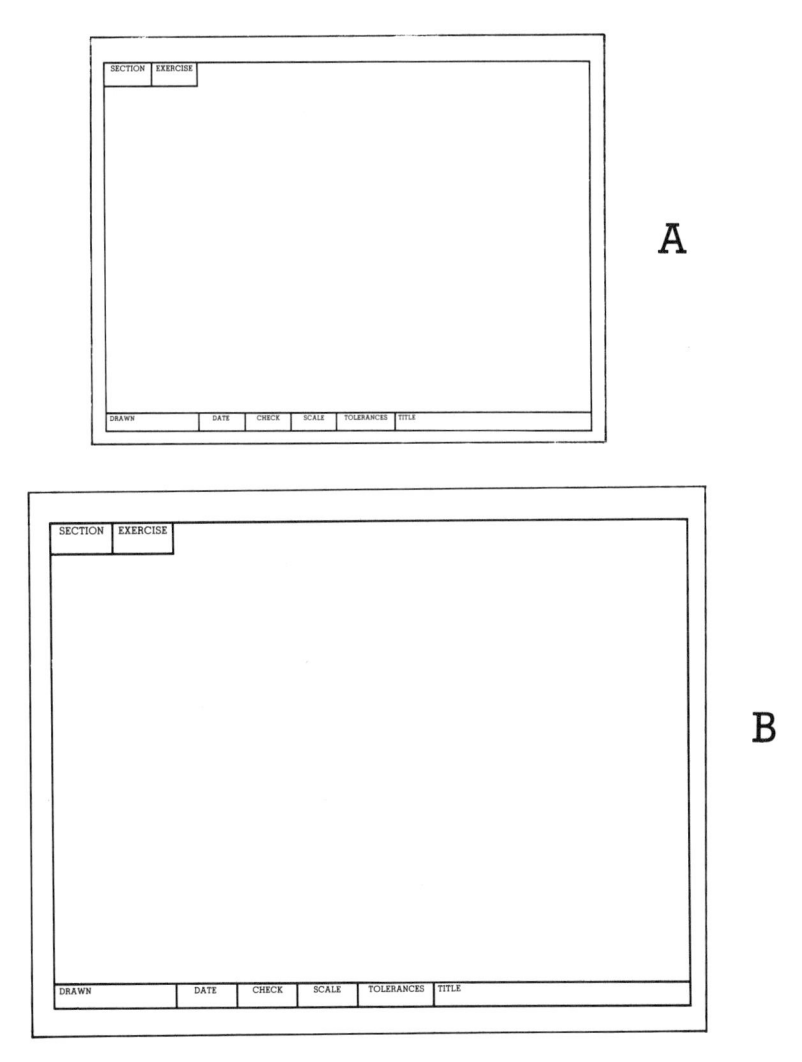

16 B. OBJECTIVE: Redraw the sketch identified below as a CAD drawing. The objects should be drawn proportional to the ones shown and centered on an "A" format. Use PLACE LINE, CIRCLE, AND RADIUS commands. The title of the drawing is OBJECTS. Include all lines.

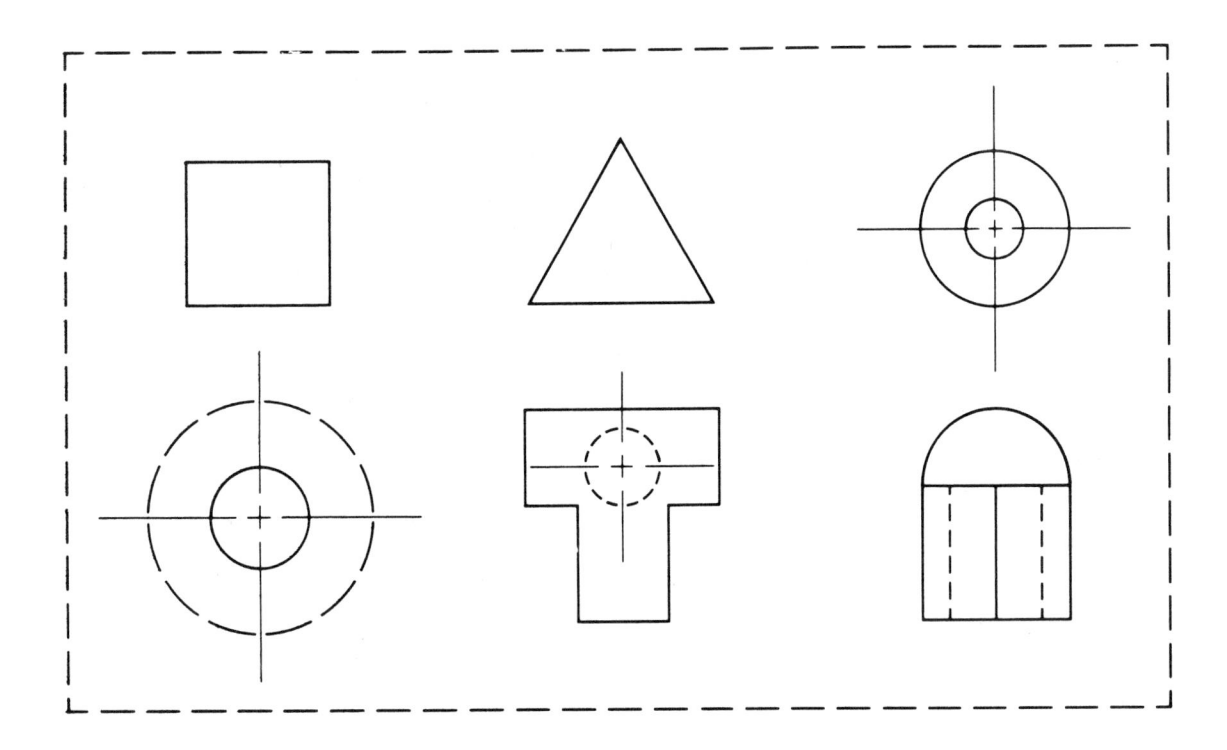

16 C. OBJECTIVE: Draw the three figures shown in column A. At completion, and after approval by the instructor, change to look like those in column B by EDITING, ERASING, and/or PLACING LINE commands. Add text information. Use "A" format. The title of the drawing is EDITING.

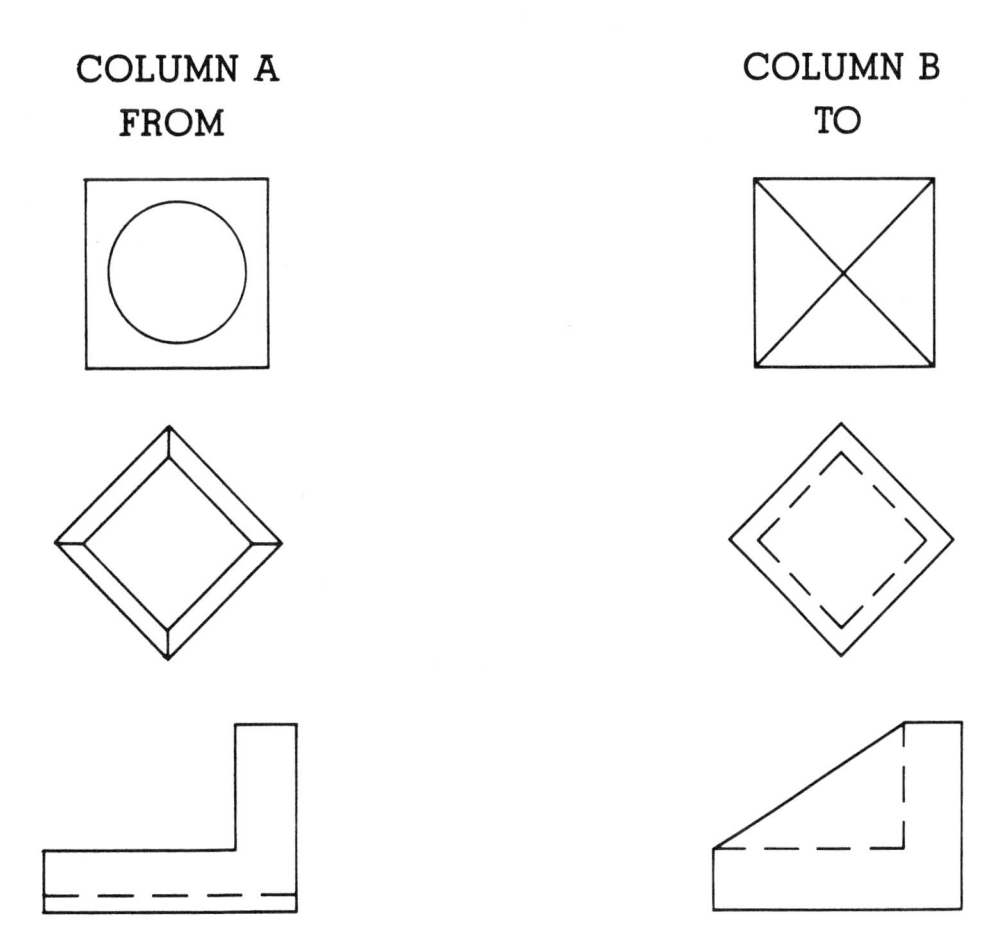

COLUMN A
FROM

COLUMN B
TO

16 D. OBJECTIVE: (a) Produce the TEMPLATE drawing shown below. Do not dimension. Use "A" format. (a) Add CROSSHATCH to the drawing.

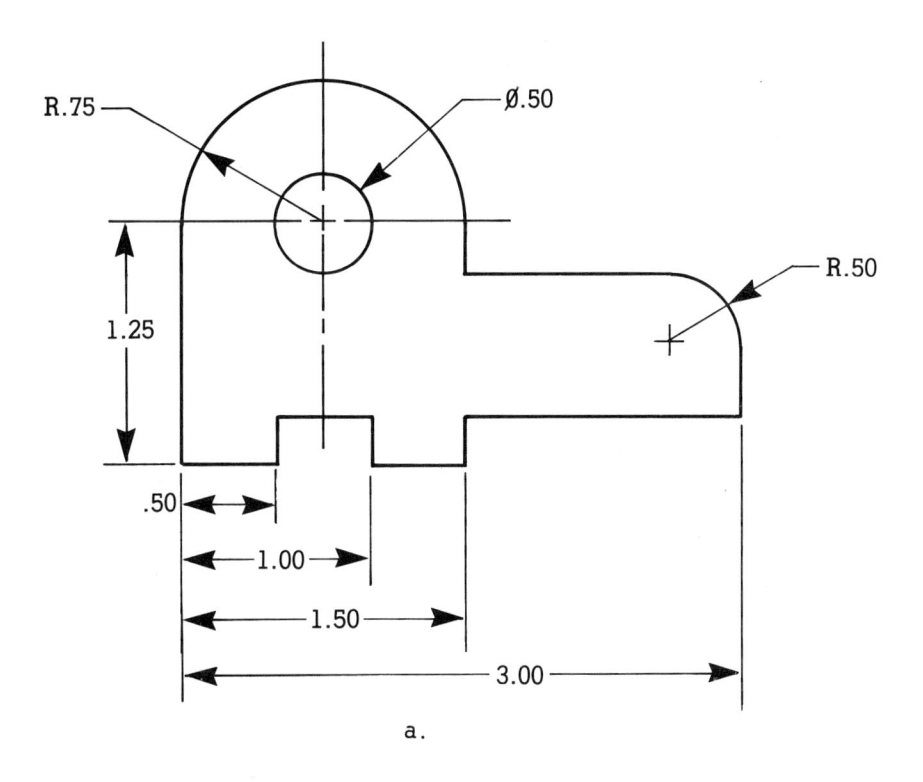

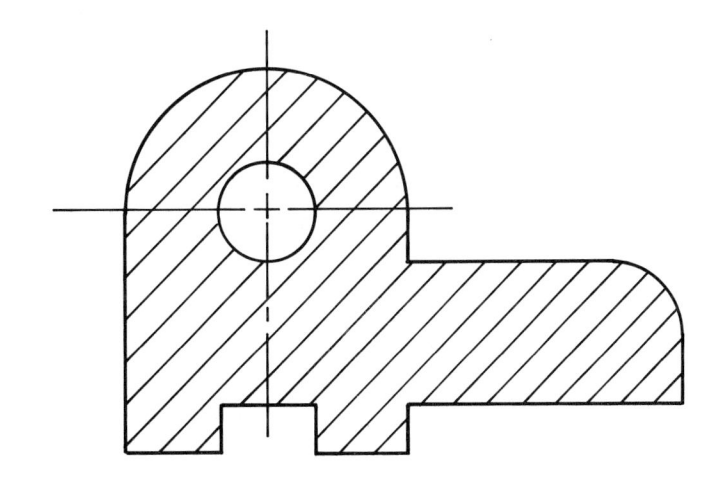

a.

b.

16 E. OBJECTIVE: Using COORDINATE DIMENSIONING practices produce the drawing below. Use "A" format. Draw full size. The title is STRAP. Include dimensions.

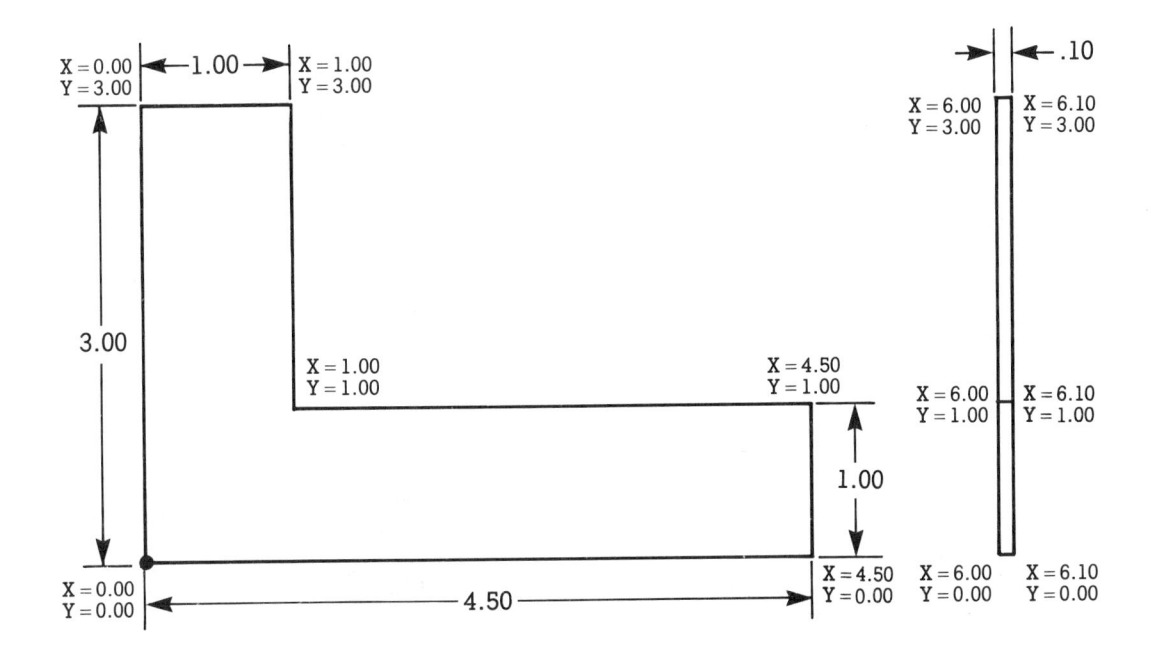

16 F. OBJECTIVE: Using either DELTA COORDINATE COMMAND or a MENU TABLET produce a CAD drawing of completed EXERCISE 2G, shown below. Draw full size. Do not add dimensions. Use "A" format.

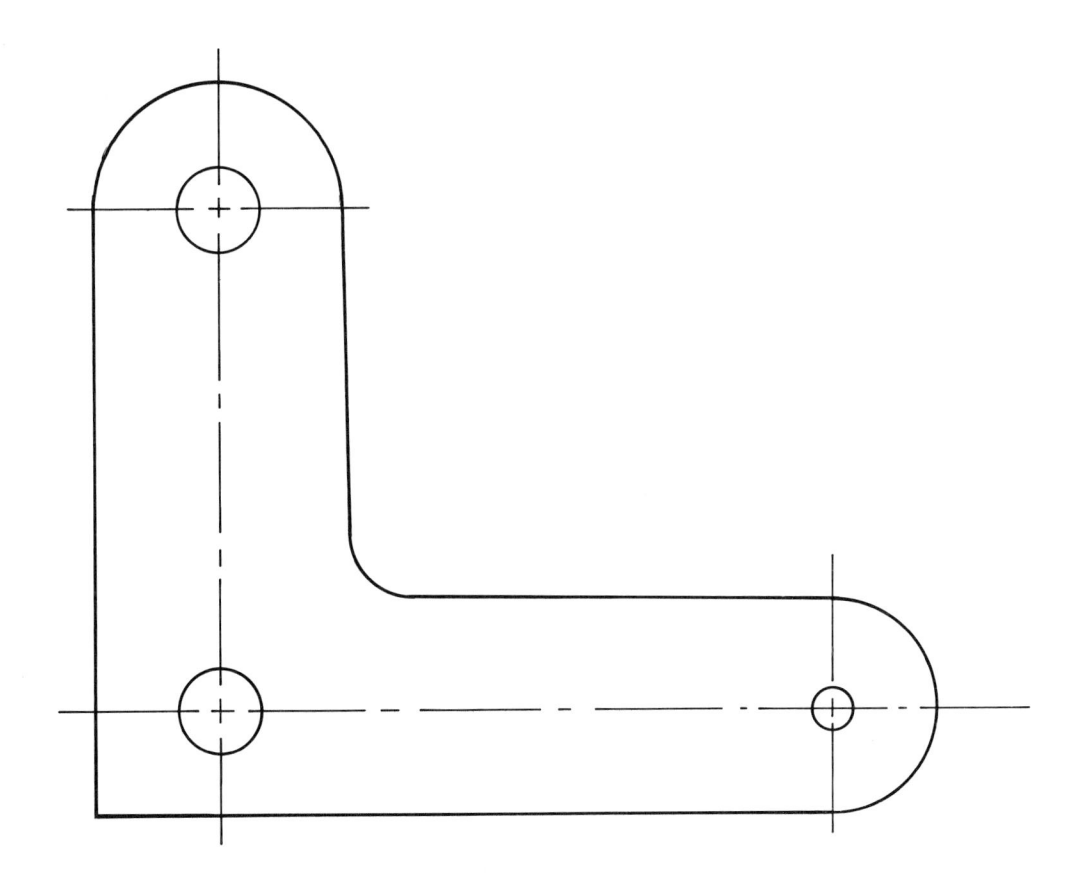

16 G. OBJECTIVE: Draw the CUBE shown below using the POLAR COORDINATE method. Follow the sequence identified to the right of the illustration.

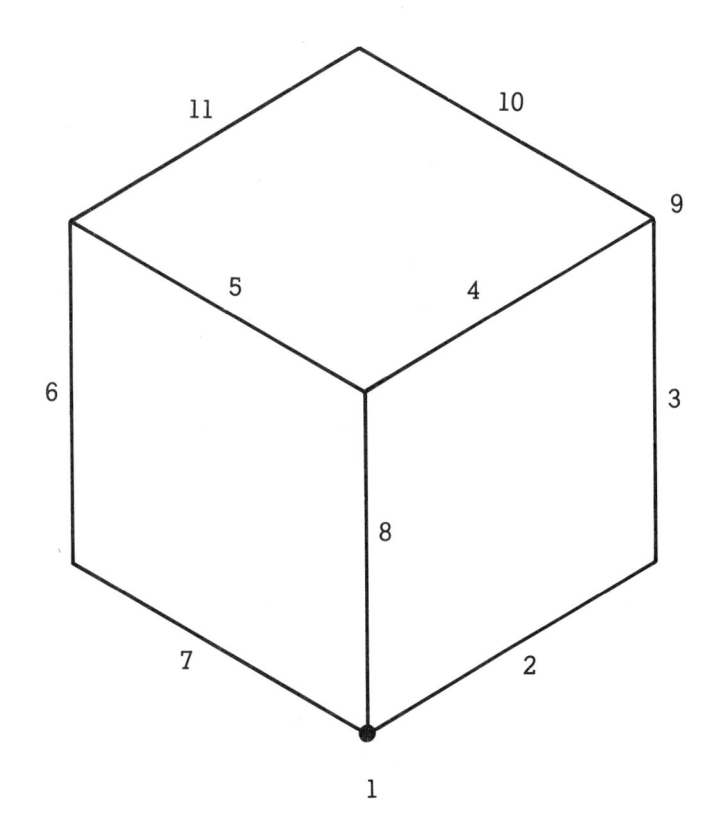

V : PA,X	RET 1
V : P30, 1;2	RET 2
V : P90, 1;2	RET 3
V : P210, 1;2	RET 4
V : P150, 1;2	RET 5
V : P-90, 1;2	RET 6
V : P-30, 1;2	RET 7
V : P90, 1;2	RET 8
V : P30, 1;1	RET 9
V : P150, 1;2	RET 10
V : P210, 1;2	RET 11

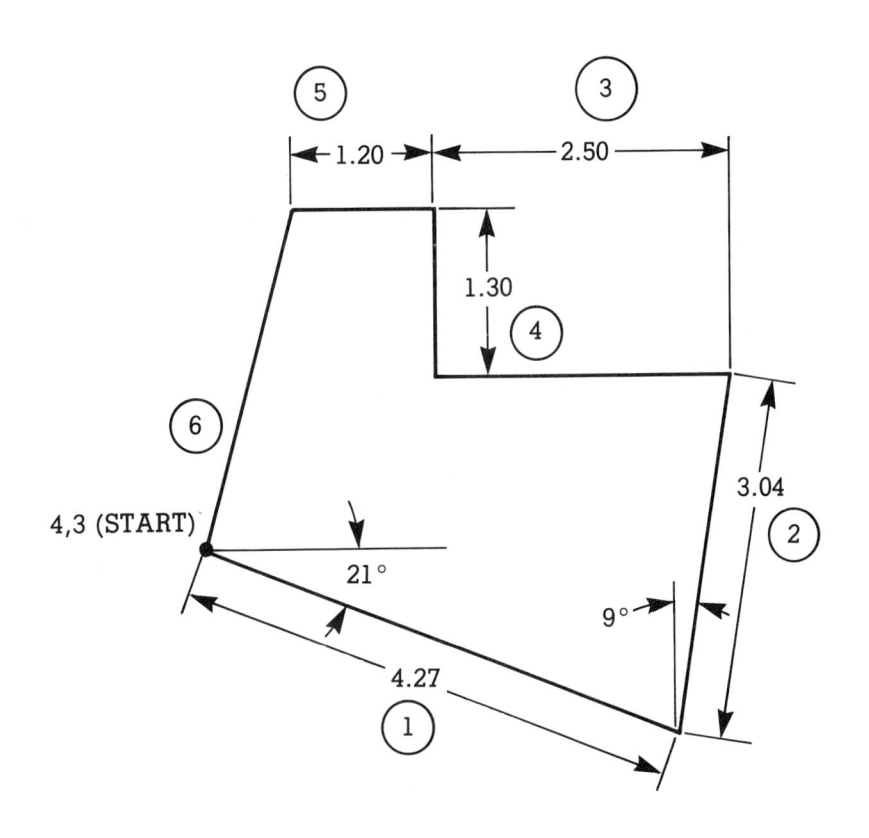

16 H. OBJECTIVE: Complete the programming statements below for producing the following drawing.

XY =	_____	(START)
DI =	_____	(LINE 1)
DI =	_____	(LINE 2)
DL =	_____	(LINE 3)
DL =	_____	(LINE 4)
DI =	_____	(LINE 5)
XY =	_____	(LINE 6)

16 I. OBJECTIVE: On "B" format produce an isometric CAD drawing of the SHOE HOUSING shown below. Draw full scale. Do not dimension.

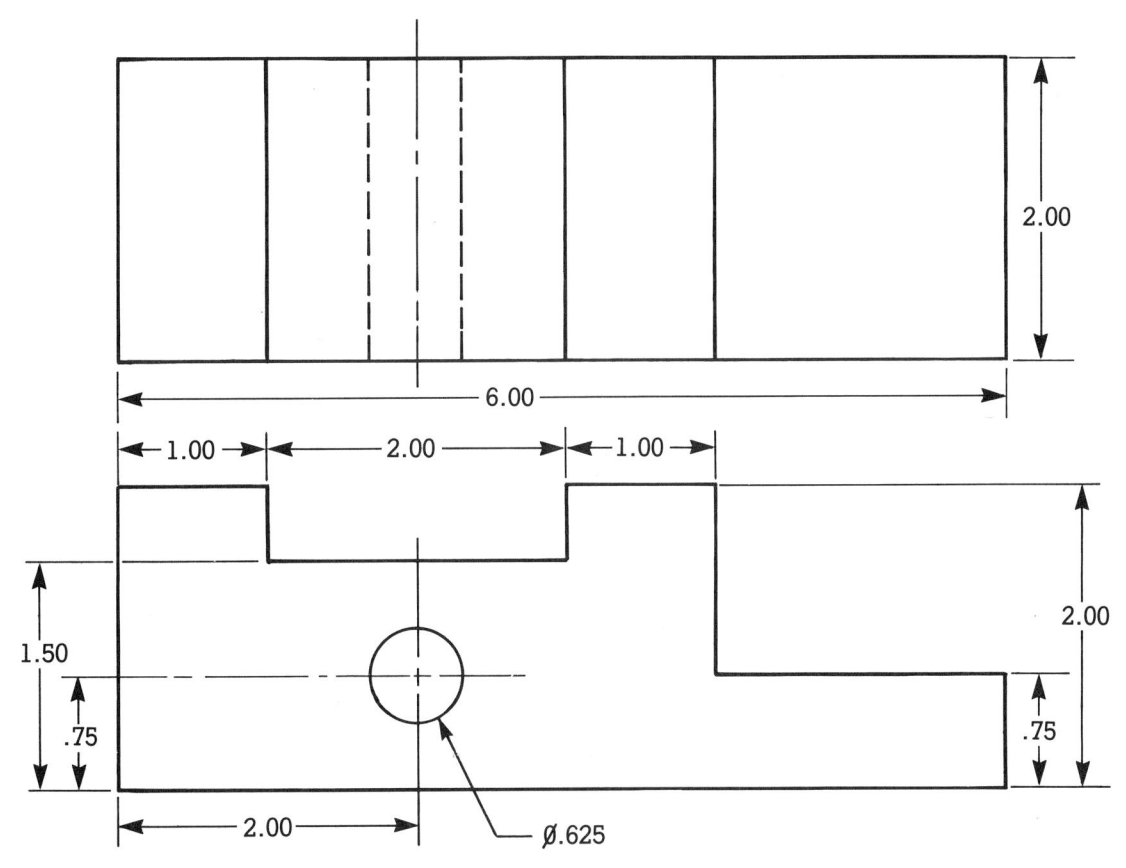

16 J. OBJECTIVE: Using "A" format produce a cabinet type oblique drawing of the V-BLOCK below. Scale is full. Do not dimension.

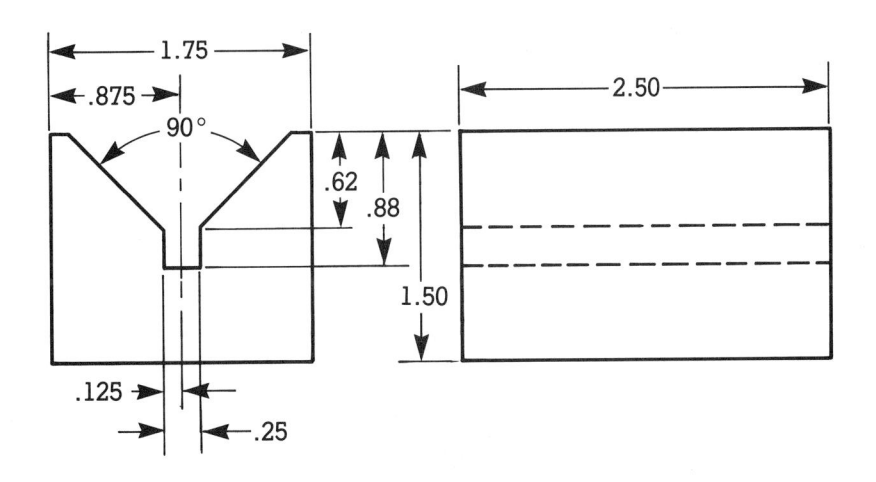

16 K. OBJECTIVE: Scale the STEPS drawing below and draw a two-vanishing point perspective drawing. Use "B" format.

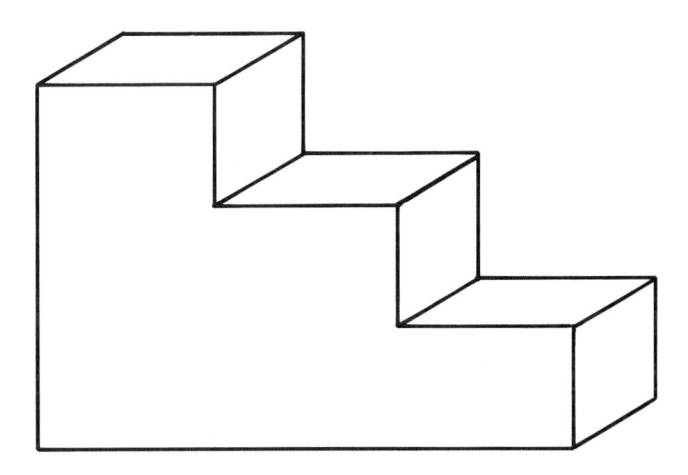

16 L. OBJECTIVE: Produce a three-view drawing of the MOUNTING BRACKET below. Use CONVENTIONAL or AUTOMATIC DIMENSIONING practices. Draw twice size. Call up format "B". Note! Dimensions are in millimeters.

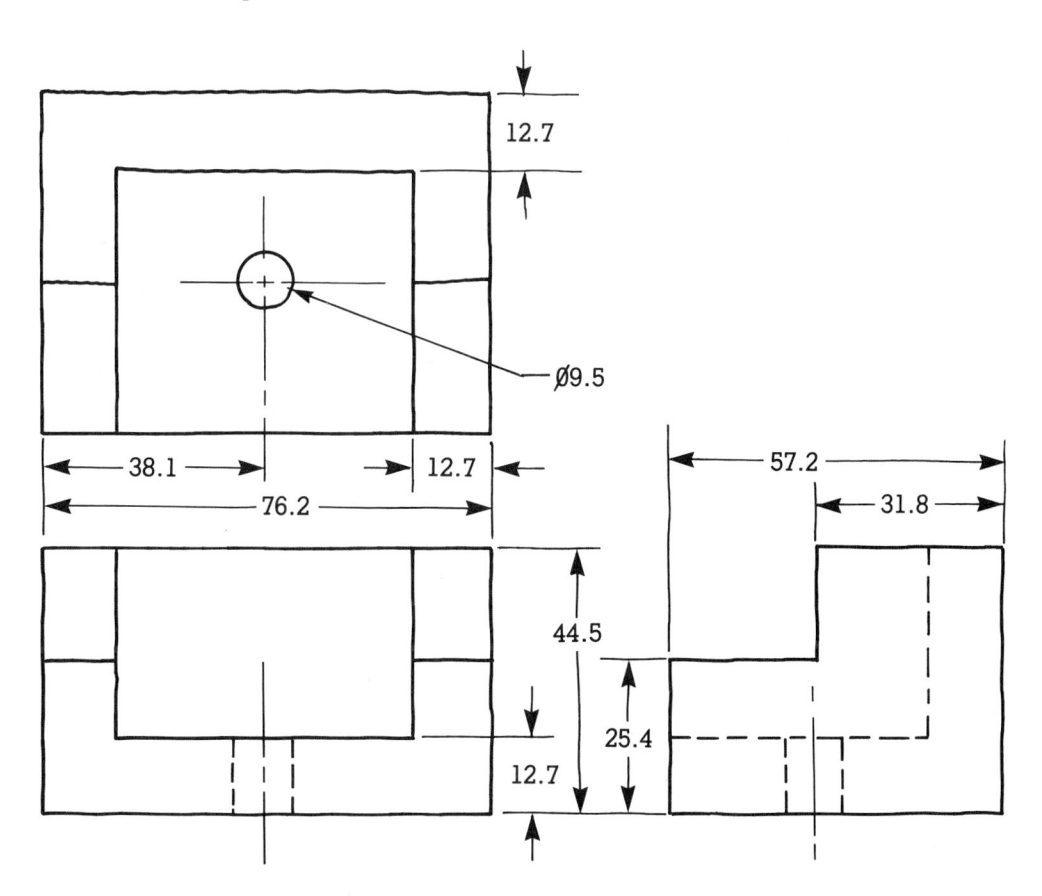

16 M. OBJECTIVE: On "A" format draw the development of the BRACKET shown below. Calculate and show the developed length.

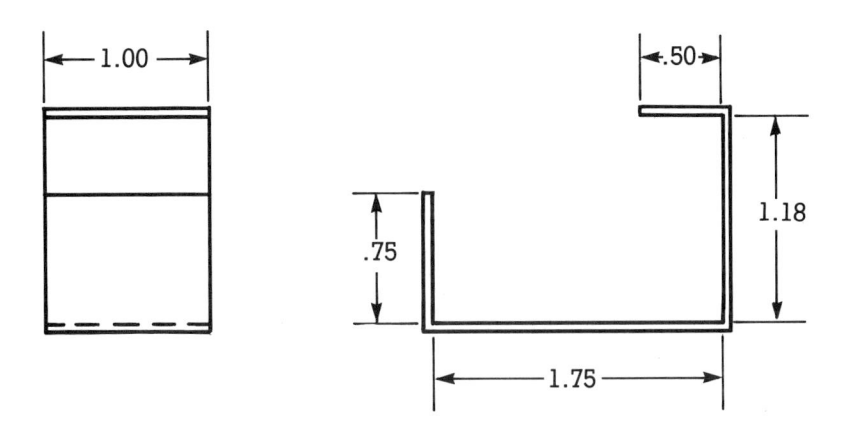

16 N. OBJECTIVE: Using "B" format determine the TRUE LENGTH OF LINE **cd** in the descriptive geometry problem below. Identify all points, lines, and projections. Scale drawing for accuracy.

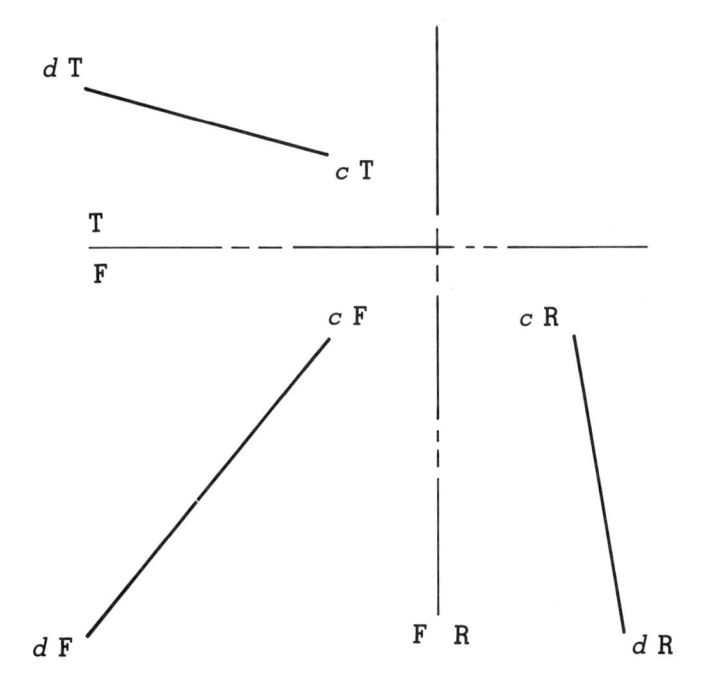

16 O. OBJECTIVE: On "B" format draw the front auxiliary view of the WEDGE shown below. Also include the front and top views. Draw full size. Note! Dimensions are in millimeters.

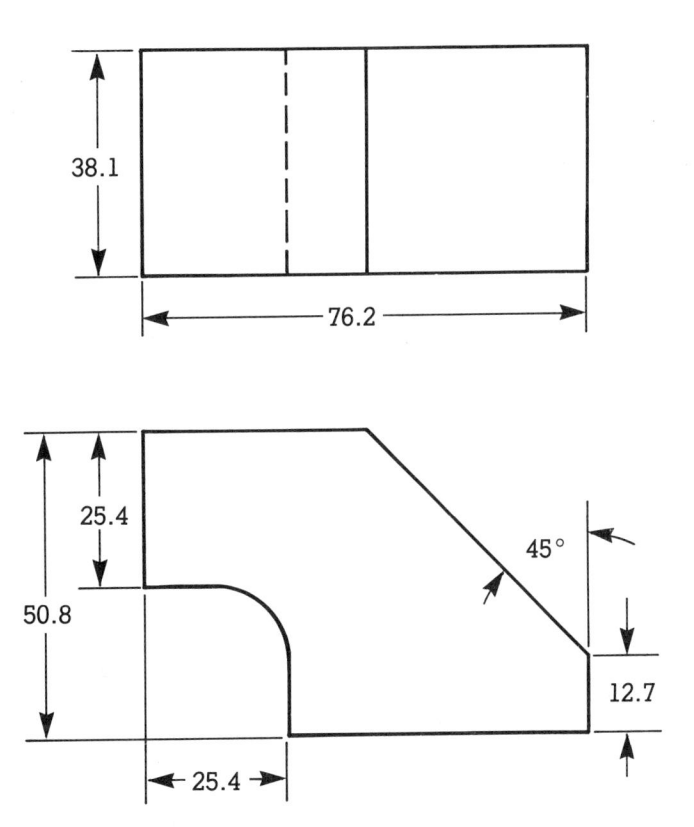

16 P. OBJECTIVE: On "B" format and using the RADIAL LINE DEVELOPMENT method construct the development of the truncated pyramid shown below. Use 6.3 wide tabs. Dimensions are metric. Title is TRUNCATED PYRAMID DEVELOPMENT. Draw full size.

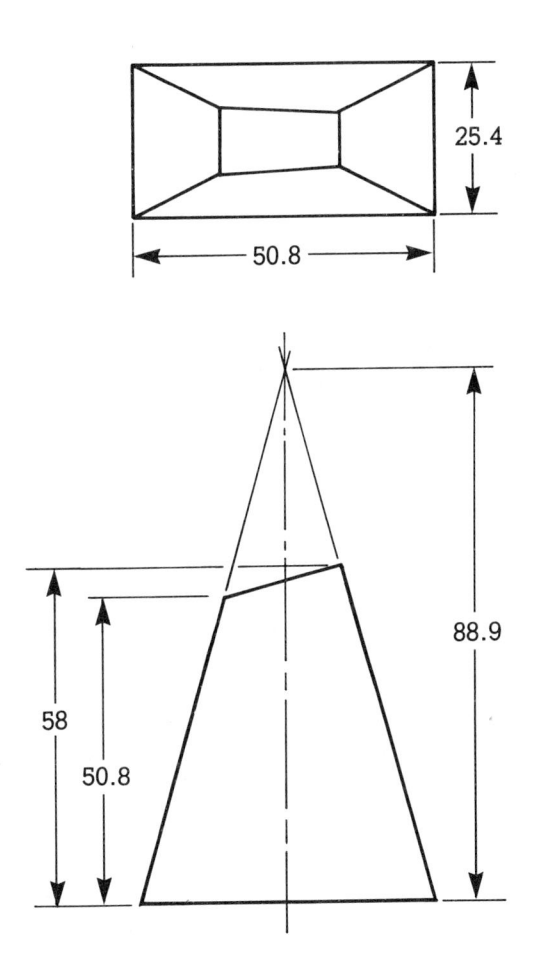

16 Q. OBJECTIVE: Draw to full scale, three (3) views of the HOUSING STRAP shown. Add all dimensions, notes, material, and finish information. Draw on "B" format.

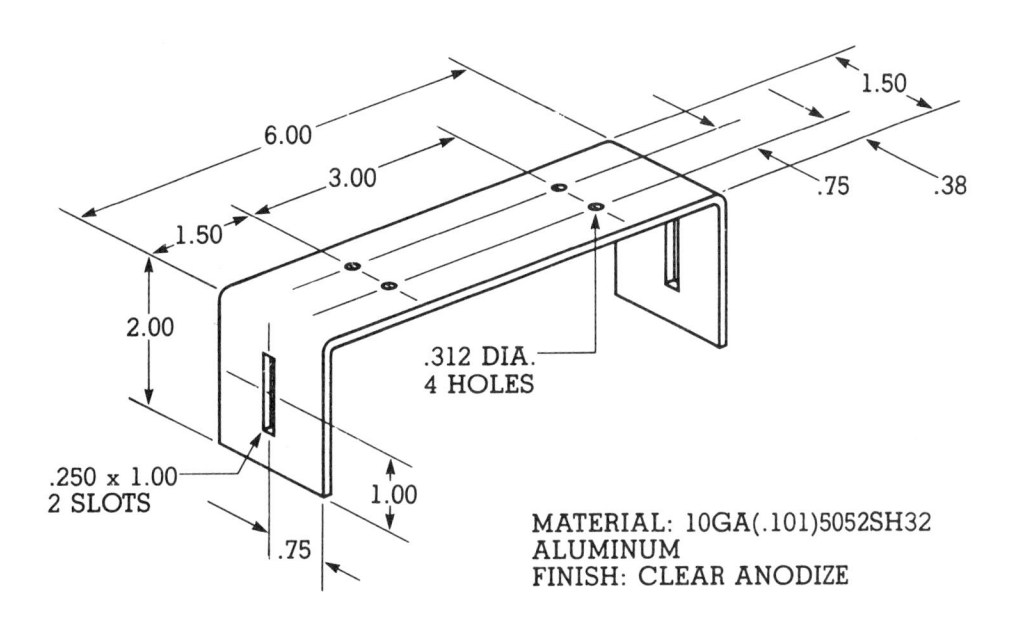

MATERIAL: 10GA(.101)5052SH32 ALUMINUM
FINISH: CLEAR ANODIZE

16 R. OBJECTIVE: Produce to a scale of 1:1 on "B" format an isometric drawing of the CHASSIS COVER shown below. Add all dimensions, notes, material, and finish information.

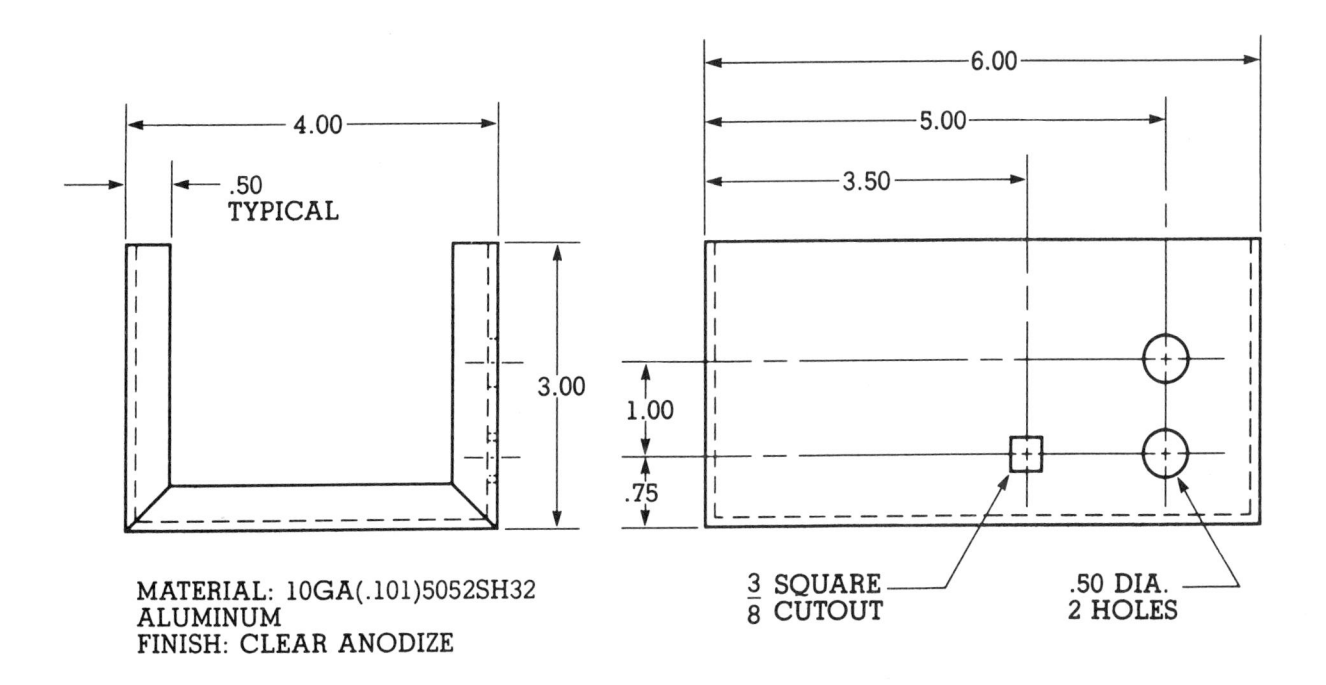

MATERIAL: 10GA(.101)5052SH32 ALUMINUM
FINISH: CLEAR ANODIZE